World Symposium on Smart Materials and Applications (WSSMA 2019)

IOP Conference Series: Materials Science and Engineering
Volume 649

Harbin, China
5 – 7 July 2019

ISBN: 978-1-5108-9846-2
ISSN: 1757-8981

Printed from e-media with permission by:

Curran Associates, Inc.
57 Morehouse Lane
Red Hook, NY 12571

Some format issues inherent in the e-media version may also appear in this print version.

Institute of Physics
Dirac House, Temple Back
Bristol BS1 6BE UK

Phone: 44 1 17 929 7481
Fax: 44 1 17 920 0979

techtracking@iop.org

Additional copies of this publication are available from:

Curran Associates, Inc.
57 Morehouse Lane
Red Hook, NY 12571 USA
Phone: 845-758-0400
Fax: 845-758-2633
Email: curran@proceedings.com
Web: www.proceedings.com

Table of contents

Volume 649

World Symposium on Smart Materials and Applications (WSSMA 2019)

5–7 July 2019, Harbin Engineering University, Harbin, China

Accepted papers received: 13 September 2019
Published online: 17 October 2019

Preface

Papers

Preface

The WSSMA 2019 is the 2019 World Symposium on Smart Materials and Applications (WSSMA 2019) took place in Harbin, China, on July 05-07, 2019.

The conference program covered invited, oral, and poster presentations from scientists working in similar areas to establish platforms for collaborative research projects in this field. This conference will bring together leaders from industry and academia to exchange and share their experiences, present research results, explore collaborations and to spark new ideas, with the aim of developing new projects and exploiting new technology in this field.

The committee of WSSMA 2019 expresses their sincere thanks to all authors for their high-quality research papers and careful presentations. All reviewers are also thanked for their careful comments and advices. Thanks are finally given to IOP Publication as well for producing this volume.

The Organizing Committee of WSSMA 2019

Committee Chair

Prof. Dongyan SHI

College of Mechanical and Electrical Engineering

Harbin Engineering University

IOP Conf. Series: Materials Science and Engineering **649** (2019) 011001 doi:10.1088/1757-899X/649/1/011001

Conference Photograph

Sponsors

Incheon Disaster Prevention Research Center (IDPRC)

Recently, various efforts to prevent and prepare are vitally needed for prevention of Disasters and calamities. So we understand the necessity for technology of disaster and we built up the Incheon Disaster Prevention Research Center (IDPRC) in Incheon National University (INU) in 1997.

Accordingly, Incheon Disaster Prevention Research Center (IDPRC) in Incheon National University has progressed of research on the prevention of disasters and calamities through the various seminars, conference and lectures. These research could be conducted cooperation with Incheon National University (INU) in various fields' structure, soil, hydraulic and environment.

Incheon Disaster Prevention Research Center (IDPRC) will try to be a leader in the disaster of industry through the various research activities and global conference.

Harbin Engineering University

The predecessor of HEU, the PLA Military Engineering Institute was established, and Senior General Chen Geng acted as the first president. Chairman Mao gave the instructions for the establishment. According to the types of the services, the Institute was subdivided into five parts: Department of Air Force Engineering, Department of Artillery Engineering, Department of Naval Force Engineering, Department of Armored Forces Engineering, and Department of Corps Engineering. HEU Intelligent Underwater Robot Lab was listed as a National Key Lab. In 2002 theMinistry of Science and Technology and Ministry of Education approved the construction of a national University Science Park within HEU. In December 2002, the Commission of Science Technology and Industry for National Defense and the People's Government of Heilongjiang Province signed an agreement to jointly construct this university. HEU was listed on the Major course pioneering platform on 2011.

IOP Conf. Series: Materials Science and Engineering **649** (2019) 011001 doi:10.1088/1757-899X/649/1/011001

Lists of Committees

Committee name	Organizer name	Organization
Honorary Chair	Dongyan SHI	Harbin Engineering University
Conference General Chairs	Jongwan. Hu	Incheon National University
Honorary Chair	Dongkeon Kim	Dong-A University
Technical Committees	J. W. Hu	Incheon National University
	J. K. Ahn	Incheon National University

Peer review statement

All papers published in this volume of *IOP Conference Series: Materials Science and Engineering* have been peer reviewed through processes administered by the proceedings Editors. Reviews were conducted by expert referees to the professional and scientific standards expected of a proceedings journal published by IOP Publishing.

IOP Conf. Series: Materials Science and Engineering **649** (2019) 012001 doi:10.1088/1757-899X/649/1/012001

Research on Performance of a Kind of Compound-type Corrosion-Inhibition Bactericide

Huang Changshan[1], Wu Jinying[1, *], Chen Min[1], Xu Huiwu[1], Zhang Min[1], Xue Mian[1] and Zhang Xiaoguang[1]

[1]Henan Academy of Sciences, Energy Research Institute Co., Ltd. Zhengzhou, China

13937140170@163.com

Abstract. In the experiment, rotative weight loss method, electrochemical polarization, MPN, etc. were used to investigate the corrosion inhibition and bactericidal performance of compound-type corrosion-inhibition bactericide composed of phosphono-, carboxyl quaternary ammonium salt, etc. on A3 carbon steel in industrial recycling cooling water. The experimental test results showed that when the content of compound agentia reached 100mg. L^{-1}, the corrosion rate of it on A3 carbon steel was 0.0366mm. a^{-1}, superior to relative standard regulations, less than 0. 075mm.a^{-1}. While metallographic-phase microscopy test showed that there is a layer of protective film on the surface of A3 carbon steel, which indicated that compound agentia had relatively better corrosion-inhibition effect on A3 carbon steel; when the content of compound agentia reached 80-100mg. L^{-1}, the sterilizing rate of compound agentia on heterotrophic bacterium and sulfate reducing bacteria were up to more than 99%, which indicated that compound agentia had good bactericidal performance.

1. Introduction

In electric power, metallurgy and chemical engineering industry, recirculating cooling water system for central air-conditioning is mainly composed of pipe and devices mainly made of carbon steel. When it is used in high-temperature circulating water with higher concentration ratio, oxygen absorption corrosion of metal and microbiological corrosion mainly caused by total general bacteria (TGB) and sulfate reducing bacteria (SRB) can be generated [1-2]. In actual application, chemical agentia is added to effectively resolve such problem. Single agentia only has single role, while different agentia mixing can decrease respective effect. Therefore, it is of great significance to studying the compound agentia with multiple functions.

The organic phosphonate with phosphono (-CPO_3H) can chelate with multiple metal ions in the water, to form the metal chelates, and its C-P bond with high chemical stability can't be easily hydrolyzed [3]; carboxyl quaternary ammonium salt has the bactericidal performance due to containing the positively-charged quaternary ammonia ions [4]. In the paper, we especially investigated the corrosion inhibition and bactericidal performance of compound-type corrosion-inhibition bactericide composed of phosphono- and carboxyl quaternary ammonium salt on recirculating cooling water system for central air-conditioning.

2. Experiment

2.1. Preparation of Compound Agentia

Compound agentia is the quaternary ammonium salt composed of phosphono (-CPO_3H) 2-

Phosphonobutane-1,2,4-tricarboxylic acid and carboxyl quaternary ammonium salt.

Preparation of phosphono organic phosphonate

Add 3 times (molar ratio) of solid chloroacetic acid into the bisphenol A hot alkaline solution of 90℃ gradually, after reflux reaction for 2h, cooling, acid pickling and filtering were performed, the filter cake was recrystallized by acetic acid, to get the 2-Phosphonobutane-1,2,4-tricarboxylic acid (PBTCA) [5] with purity of 98%.

Preparation of Carboxyl Quaternary Ammonium Salt

Weigh and add certain amount of tetradecyl methyl carbonate trimethyl quaternary ammonium salt and solvent ethyl alcohol into a 250ml four-neck flask, heat it under continuous agitating to the set temperature of 130℃, after completely dissolved, certain volume of acetic acid was added through a pressure equalizing dropping funnel. At this temperature, agitating was done for 2-3h, then the pale yellow liquid was got. After decompressed and solvent was removed in a rotary evaporator, the crude product was got. The crude product was recrystallized by ethyl acetate for 5 times, and dried for 6-8h in vacuum circumstance, then, the white solid product tetradecyl trimethyl ammonium acetate (TTAA) [6] with purity of 98% was got.

Formula of Compound-type Corrosion-inhibition Bactericide

Benzotriazole (BTA) 10 mg. L^{-1} + 30 mg. L^{-1} polymaleic acid (PMA) + 50 mg. L^{-1} 2-Phosphonobutane-1,2,4-tricarboxylic acid (PBTCA) + 500 mg. L^{-1} tetradecyl trimethyl ammonium acetate (TTAA).

2.2. Corrosion-inhibition Experiment

Corrosion inhibition of compound agentia was tested by weight loss method. Instrument: RCC-l rotary coupon corrosion test instrument, material: A3 carbon steel standard test piece, specification: 50mm×25 mm×2mm, precision: 0.1mm, hanging hole: φ4.0±0.1mm, surface area: 28cm^2; experiment medium: recirculating cooling water system for central air-conditioning indicated in table 1 of Item 1.6, test piece was completely immersed, experiment temperature: 50±2℃, experiment period: 72h. After experiment, test piece surface was cleaned with a scrubbing brush, and washed with acid, tap water, alkali, deionized water and absolute ethyl alcohol successively, dried to constant weight, and placed for 24h, then weighed, to calculate the corrosion rate [7].

2.3. Electrochemical Test

Electrochemical test was performed to investigate the corrosion inhibition of compound agentia. Instrument: CS dual unit electrochemical workstation, three-electrode system: auxiliary electrode is platinum electrode, reference is electrode saturated calomel electrode, research electrode is small cylinder of surface 1 cm^2 made of A3 carbon steel, the non-working surface is sealed with epoxy resin. Experiment medium is recirculating cooling water system for central air-conditioning, experiment temperature: room temperature.

2.4. Mirror Test

The blank sample solution and A3 carbon steel test piece in 80 mg/L compound agentia were longitudinally cut, inlaid, polished and eroded, thickness of film of A3 carbon steel test piece surface was tested in accordance with operation requirements of metallurgical microscopy and test requirements of this experiment. Test was done with a AE2000MET digital inverted metallurgical microscopy, with magnification of 400X.

2.5. Bactericidal Experiment

TGB and SRB in recirculating water at production site were used for enrichment culture of strains, bacteria quantity: TGB of 10^5～10^7个/ml; SRB of 10^2～10^6/ml. According to GB/T 22595-2008[8] and GB/T 14643.5-2009[9], number of TGB and SRB in recirculating cooling water before and after adding compound agentia was determined respectively.

2.6. Experiment Medium

The recirculating cooling water for central air-conditioning of *Crowne Plaza* was used as the experiment medium, its main parameters are as shown in table 1:

Table 1. Main Parameters of Recirculating Cooling Water (mg. L^{-1}).

Parameter Index	Ca^{2+}	Hardness	Alkalinity	Cl^-	Fe^{2+}	pH
Parameter data	108	360	203	40	0.5	7.3

3. Results and Discussions

3.1. Corrosion Inhibition of Compound Agentia

Corrosion inhibition results of different-concentration compound agentia on A3 carbon steel in recirculating cooling water refer to table 2.

Table 2. Corrosion Inhibition Role of Compound Agentia on A3 Carbon Steel in Recirculating Cooling Water.

Concentration of compound agentia (mg.L^{-1})	Corrosion rate (g.m^{-2}.h^{-1})	Corrosion rate (mm.a^{-1})	Inhibition efficiency(%)
0	0.2167	0.2434	-
20	0.1426	0.1602	34.2
40	0.1281	0.1439	37.4
60	0.0979	0.1099	54.8
80	0.0747	0.0839	65.5
100	0.0326	0.0366	85.0

Note: temperature is 50℃, soak period is 72h.

Seen from table 2: the inhibition efficiency of compound agentia increases along with increase of its concentration, its corrosion inhibition is especially apparent. When its concentration reaches 100 mg. L^{-1}, the inhibition efficiency is 85%, when the corrosion rate of A3 carbon steel reduces to 0.0366mm/a, far below the value less than 0.075mm. a^{-1} specified in GB/T 50050-2017 Code for *Design of Industrial Recirculating Cooling Water Treatment* [10]. This indicated that the compound agentia with the structure of phosphono- and carboxyl quaternary ammonium salt has relatively better corrosion-inhibition effect on A3 carbon steel.

3.2. Electrochemical Behavior of Compound Agentia

3.2.1. Polarization Curve of Different-concentration Compound Agentia

Potentiodynamic scan curve of A3 carbon steel in recirculating cooling water with concentration-compound agentia refer to figure 1.

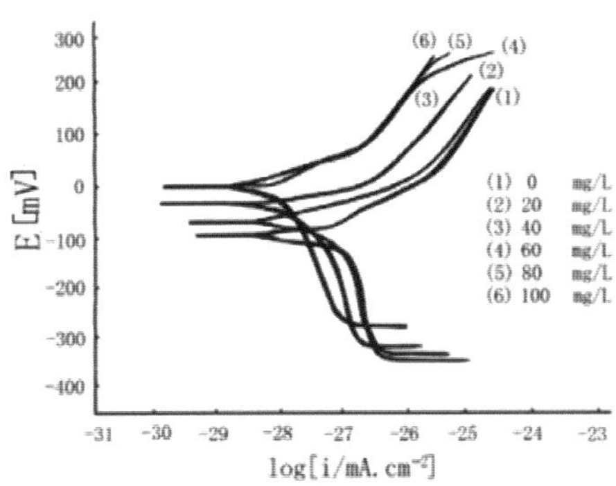

Figure 1. Polarization Curve of A3 Carbon Steel in Different-concentration Compound Agentia Solution.

Seen from figure 1: cathodic polarization and anodic polarization of polarization curve of A3 carbon steel in recirculating cooling water with compound agentia apparently increase, indicating that the compound agentia can apparently inhibit the cathodic and anodic polarization process of electrode, and belongs to the mixed-type corrosion inhibitor. The lower-concentration compound agentia can mainly inhibit the cathodic polarization process, while high-concentration one is of higher effect in inhibiting the anodic polarization process.

3.2.2. Corrosion Potential of Different-concentration Compound Agentia

Corrosion potential of A3 carbon steel in recirculating cooling water with different-concentration compound agentia refers to table 3.

Table 3. Corrosion Potential of Different-concentration Compound Agentia on A3 Carbon Steel.

Concentration of compound agentia($mg.L^{-1}$)	0	20	40	60	80	100
E(mV)	-653	-517	-501	-485	-469	-452

Seen from table 3, it can be seen that the compound agentia can make the corrosion potential apparently positively shifted, when its concentration is up to 100 mg. L^{-1}, positively shifted potential is about 200 mV. By comprehensively analyzing such electrochemical experiment data and weight loss experiment results, we find that the significant positive shift of corrosion potential and increased cathodic/anodic polarization can greatly decrease the corrosion rate of A3 carbon steel in industrial recycling cooling water.

3.2.3. Surface Appearance

figure 2 refers to the appearances of blank A3 carbon steel coupon not provided with compound agentia and A3 carbon steel coupon provided with compound agentia. It can be seen that the surface of blank A3 carbon steel coupon not provided with compound agentia is in coarse and dark black color. While the surface of A3 carbon steel coupon provided with compound agentia of 80 mg/L is in original carbon steel metal luster.

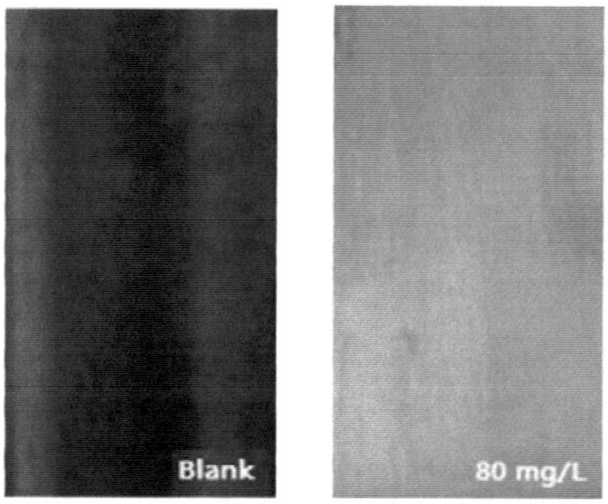

Figure 2. Surface Appearances of Blank A3 Carbon Steel Coupon Not Provided with Compound Agentia and Couponprovided with Compound Agentia of 80mg. $L^{-1.}$

3.2.4. Test on Surface Film

figure 3 and 4 refer to the metallographs of cross sections of blank A3 carbon steel coupon and A3 carbon steel coupon provided with compound agentia of 80 mg. L^{-1}.

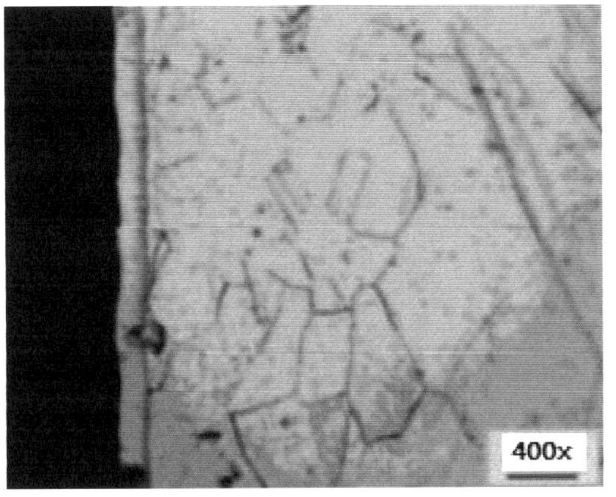

Figure 3. Metallograph of A3 Carbon Steel Coupon in Blank Sample Solution.

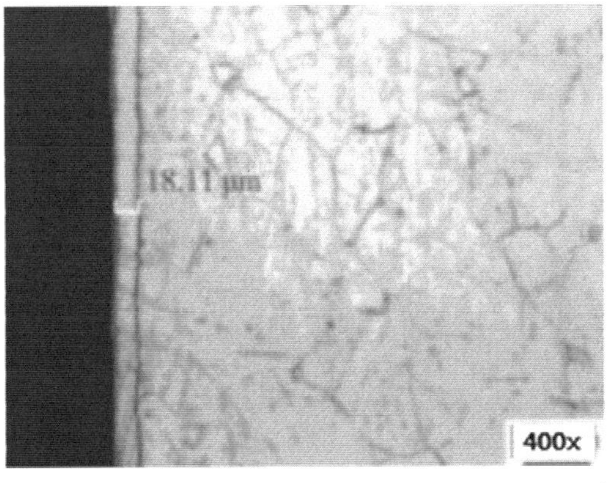

Figure 4. Metallograph of A3 Carbon Steel Coupon in 80 mg.L^{-1} Sample Solution.

From the metallographic microstructure test results of A3 coupon cross section, it can be seen that surface of A3 carbon steel coupon in blank sample solution (figure 3) has no film, its appearance is dark and coarse. A3 carbon steel in compound agentia sample solution has a layer of even and compact film with uniform thickness of about 18.11μm. After treated by compound agentia, the film is in bright color and metal luster, apparently, its corrosion resistance is greatly enhanced than that not in the compound agentia sample solution.

3.3. Bactericidal Performance of Compound Agentia

Experimental strains are TGB and SRB enriched-cultured in recirculating cooling water at the production site. Experimental results were got by the principle of dilution-to-extinction method. The bactericidal experiment results of compound agentia refer to table 4 and table 5.

Table 4. Rate of Sterilizing TGB after Adding Compound Agentia for 24h.

Compound agentia(mg.L^{-1})	80	100
Survived bacteria(pieces/ml)	1.5×10^3	0.2×10^2
Sterilizing rate(%)	99.4	99.992

Note: initial bacteria number is 2.5×10^5/ml.

Table 5. Rate of Sterilizing SRB after Adding Compound Agentia for 24h.

Compound agentia($mg.L^{-1}$)	80	100
Survived bacteria(pieces/ml)	1.5×10^3	0.3×10^2
Sterilizing rate(%)	99.75	99.995

Note: initial bacteria number is 6.0×10^5/ml.

The experimental results of table 4 and 5 indicated that after the compound agentia acts for 24h, its sterilizing rates on TGB and SRB are up to more than 99%. The compound agentia is of the molecular structure with positively-charged quaternary ammonium cations, which can change the osmotic pressure of cell wall by being absorbed onto the negatively charged bacteria surface, making the thalli ruptured. In addition, hydrophobic group and hydrophilic group in compound agentia can penetrate into the lipoid layer and protein of bacteria cell, to result in enzyme inactivation, protein denaturation and death [4], thus, the compound agentia has good bactericidal performance.

4. Analysis on Principle of Corrosion-Inhibition Bactericide

Compound agentia contains phosphono- and carboxyl quaternary ammonium cation and hydrophobic alkyl long-chain group, its synergistic effect can realize the corrosion inhibition on metal carbon steel and sterilizing effect on TGB and SRB. Phosphonyl functional groups ($-CPO_3H$) can chelate with Fe^{2+}, Ca^{2+}, Mg^{2+} and other ions [11] in recirculating cooling water, then, the generated chelates can cover the surface of carbon steel metal. The positive charges carried by ammonium cation carboxylate group can be absorbed at the cathode region of metal carbon steel surface by the electrostatic attraction. Additionally, the non-polar hydrophobic alkyl long-chain group can be vertical to metal carbon steel surface and stretched into the solution, then, the formed water-repellent layer can hinder contact of solution with carbon steel surface. Thus, under the joint action of functional groups in compound agentia, a layer of complete compound protective film can be formed on the carbon steel surface, to inhibit anodic stripping of carbon steel and cathode diffusion of dissolved oxygen in the water, to realize better corrosion inhibition. The weight loss test indicated that when the concentration of compound agentia reaches 80-100 mg/L in recirculating cooling water, the surface of carbon steel coupon is in even and apparent metal luster, without rustiness. The appearance photos and cross section metallographic tissue test experiment performed on blank A3 carbon steel experiment coupon not in the compound agentia and A3 carbon steel experiment coupon in 80 mg/L compound agentia also verified such viewpoint.

5. Conclusion

In this research, the weight loss and electrochemical test results indicated that the compound-type corrosion-inhibition bactericide can inhibit the cathodic and anodic reactions during A3 carbon steel corrosion, corrosion rate on A3 carbon steel was up to 0.0366 mm/a, inhibition efficiency was up to 85 %, superior to the corrosion indexes specified in relevant standards. The metallographic test on surface film also indicated that the compound agentia can form a layer of compact and even film on A3 carbon steel surface, to greatly enhance the anti-corrosion performance of carbon steel. The bactericidal test results indicated that sterilizing rate of compound-type corrosion-inhibition bactericide on SRB and TGB were up to more than 99%.

The compound-type corrosion-inhibition bactericide has better corrosion inhibition and bactericidal performance. It can be widely applied to corrosion resistance and sterilization on recirculating cooling water system pipe network and carbon steel equipment of electric power, chemical engineering industry, metallurgy and central air-conditioning.

References

[1] Dong Q Y and Zhang Q 2004 *Corrosion and Protection* Synergistic Effect of Carboxylic Acid Copper Corrosion Inhibitor and Benzotriazole in Seawater **25** 426-28.

[2] Li Y 2015 *Cleaning World* Corrosion Properties and Corrosion Inhibition Mechanism of Bis-quaternary Ammonium Compound Inhibitors **31** 24-9.
[3] Kalman E, Karman F H, et al. 1993 *Corros. Sci.* Inhibition efficiency of N-containing carboxylic and carboxy-phosphonic acids **35** 1477-82.
[4] Zhang S C, Wu Z L and Chen J P 2001 *Industrial Water Treatment* Researches on the Performance and Corrosion Inhibiting Mechanism of Corrosion Inhibitor and Germicide, PS-12 **21** 16-8.
[5] Jia F C 2008 *J. Henan Mech. Electric. Eng. College* Performance Measurement of DG4 Corrosion and Scale Inhibitor and Bactericide **16** 10-1.
[6] Yan H C 2012 Synthesis and Properties of Quaternary Ammonium Salts with Carboxylate Counterions, China Research Institute of Daily Chemical Industry.
[7] GB/T 18175-2014 2014 Determination of Corrosion Inhibition Performance of Water Treatment Agents - Rotation Specimen Method.
[8] GB/T 22595-2008 2008 Test Method for Efficacy of Antimicrobials - Aerobic Bacteria [S].
[9] GB/T 14643.5-2009, Industrial Circulating Cooling Water - Sulfate - Reducing Bacteria - MPN Test. 2009.
[10] GB/T 50050-2017, Code for Design of Industrial Recirculating Cooling Water Treatment. 2017.
[11] Ma W J and Tang L Q, et al. 2016 *Corrosion Protection Petrochem. Industry Res.* Progress of Bactericide with Corrosion Inhibition Performance **13** 1-5.

Development of Low Cost Lightweight Aluminum Foam for Railway Transportation

J W Li[1, a*], C F Shan[1, b], Y L Teng[1, c] and Z C Gai[2, d]

[1]Department of Materials Manufacturing and Processing,
School of Materials Science & Engineering, Dalian Jiao-Tong University,
794 Yellow River Rd., Dalian 116028, China
[2]Shenyang Railway Experimental High School, 31 Nanjingnanjie, Heping District,
Shenyang, 110005, China

joelee0527@163.com, 549730522@qq.com, tengyingli66@163.com,
1067258730@qq.com

Abstract. Metallic foams are commonly produced using hydride and carbonates foaming agents. However, carbonate foaming agents are safer to handle than hydrides and produce aluminum foam with a fine, homogenous cell structure, low cost and easily available. The number of pores per inch (PPI) and relative density of the foam play an important role on their physical and mechanical properties. Hence it is very important to investigate effect of shell foaming agent on PPI and relative density. The present work deals with the effect of additive amount of the shell forming agent on the physical properties of an AlSi7Mg alloy closed cell foam. The foam was produced with different additive amount of shell (1, 3, 5 wt %) as a foaming agent. The PPI and density of the foam produced with different additive amount of shell as foaming agent are determined. Relative density is in the range of 0.21-0.29, PPI are in the range of 11-19 for the produced AlSi7Mg alloy closed cell foam. Therefore, the appropriate addition amount of shell is 3 % by weight for production of a uniform cell structure from the viewpoints of efficiency and economy.

1. Introduction

Metallic foams have lately attracted considerable attention as a lightweight structural components and energy absorption parts in automobile, railway transportation and aerospace industries. Metal can be foamed in many methods [1, 2].

Closed cell foams can be produced by (a) bubbling gas through molten Al-SiC or Al-Al_2O_3 alloys; (b) by stirring a foaming agent (typically TiH_2) into a molten alloy; (c) consolidation of a metal powder with a particulate foaming agent (TiH_2) followed by heating into the mushy state when the foaming agent releases hydrogen, expanding the material [3].

Aluminium foams synthesized with $CaCO_3$ powder as a blowing agent can be prepared by both powder metallurgy and melt route. The properties of aluminium foam significantly depend on its porosity, so that a desired property can be tailored by controlling the foam density [4]. The foams can be produced by $CaCO_3$ having appreciably finer cells and more uniform cell structures than currently available melt route foams. The decomposition characteristics of $CaCO_3$ is well suited to foaming of aluminium melts [5]. An addition of $CaCO_3$ into the molten metal alloy in an amount ranging from about 0.5 % to about 4.0 wt% is sufficient to provide sufficient metal oxide phases to stabilize liquid metal foam. The reactive gas producing particle is selected from the group consisting of $CaCO_3$, magnesium carbonate, magnesium-shell (dolomite) as a foaming agent [6].

Dalian, as a coastal city, is rich in shellfish, however, most of the shells are discarded as garbage. In the present work AlSi7Mg closed cell foam has been produced using zero cost shell as a foaming agent by melt route. From an application point of view, the main concern is to control the size of the pore, porosity level and foam structure, which affect the physical and mechanical properties of the foam. Hence, in this work, effect of particle size of shell on porosity, pores per inch and relative density were studied.

2. Materials and Methods

2.1. Materials

AlSi7Mg cast aluminium alloy was used as a base material. The reinforcement phase consisted of MnO_2 particles with purity of 99.0 wt % and mean mass particle size of 45 μm. Heating MnO_2 particles for 1 h at 300°C and then for 2 h at 150°C in a conventional air furnace was carried out to improve the wettability between MnO_2 and AlSi7Mg melt by removing adsorbed gases from the surface of particles. Oyster-shell powders were used as a foaming agent, shown in figure 1 respectively. Shell powders were also heat treated at 200°C for 2 h to remove humidity and adsorbed gases from the surface and improve wetting properties and dispersion of shell powders in molten metal. Scanning Electron Microscope (SEM) microphotographs of heat treated shell powders are respectively shown in figure 2.

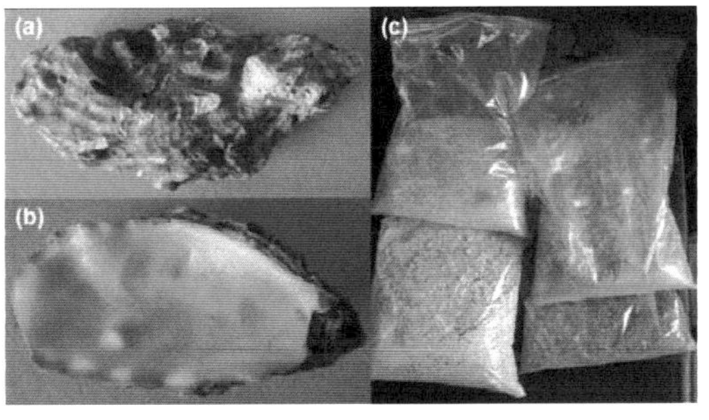

Figure 1. the oyster-shell (a), (b) and its powders (c) were used as a foaming agent.

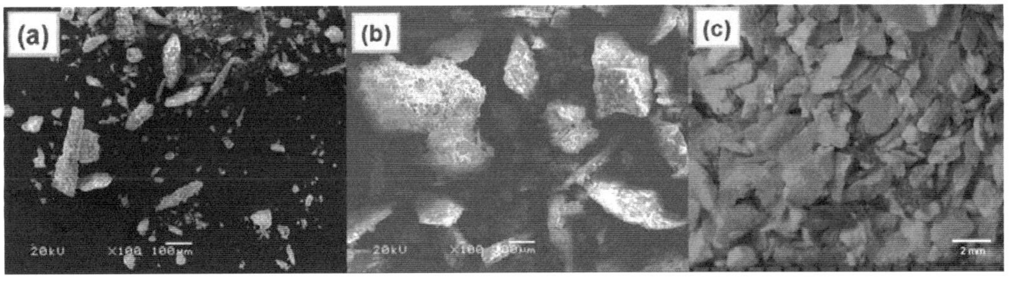

Figure 2. Microphotographs of oyster shell powders were used as a foaming agent.

(a) ≤150μm; (b) ≥300μm; (c) ≥850μm.

2.2. Methods

Required quantity of AlSi7Mg alloy was weighed and melted in a graphite-clay crucible by using an electric resistance furnace. Molten alloy temperature is measured using thermocouple. When the temperature reaches around 670 °C, furnace was switched off and melt was stirred with stainless steel stirrer for about 60 s. Required particle size of shell of 1-3-5 % by weight was sieved, preheated and added into the molten metal while stirring. The stirring was continued till the foaming agent was distributed uniformly. The shell decomposes and releases the gas when it was directly in contact with molten AlSi7Mg alloy. The crucible was kept about 30 min in the furnace until it completes the foaming process. figure 3(a) and (b) shows the stirring and foaming process for the oyster-shell powders. Then the crucible was taken out from the furnace, the foam was cooled in ambient air.

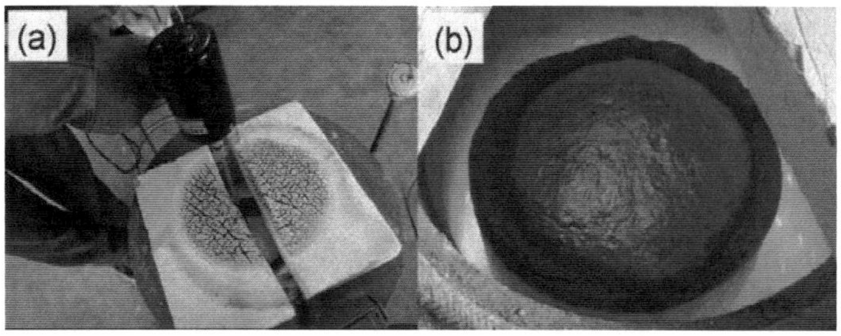

Figure 3. The stirring (a) and foaming process (b) for the oyster-shell powders.

3. Results and discussion

Figure 4 shows the influence of shell addition on the cell structure of foamed aluminum. A comparison of the cell structure shows that a shell addition of 1 wt% also creates a bubble-free zone at the bottom due to the presence of gravitational drainage effect [7]. In all cases, solid (unfoamed) portion was present in the bottom region of the foam ingot. Similar observations have been reported in the past and were attributed to liquid drainage [8]. The shell addition not only has an effect on the content of carbon dioxide gas release, but also has a relation with the melt viscosity. Two neighboring bubbles in the melt are separated by a liquid film in the liquid foam. The liquid in the film tends to flow due to surface tension [9]. If the melt with enough carbon dioxide content and a sufficient viscosity, minimizes the flow of the film the liquid drainage may be reduced [10]. This implies that the shell addition of 1 wt% is insufficient. So, the less structural stability consisting of a few irregular cells and solid aluminum phase were observed (figure 4(a)). On the contrary, a shell addition of 3.0 wt% induces a greater uniformity of spherical cell distribution (figure 4(b)). This results in the production of good foams with higher porosity. However, a shell addition of 5.0 wt% in figure 4(c) produced a partial combination of the cells. If the released gas was fully consumed to foam the specimen, the volume of the specimen will increase remarkably. However, experimental results have shown that the attainable volume is limited.

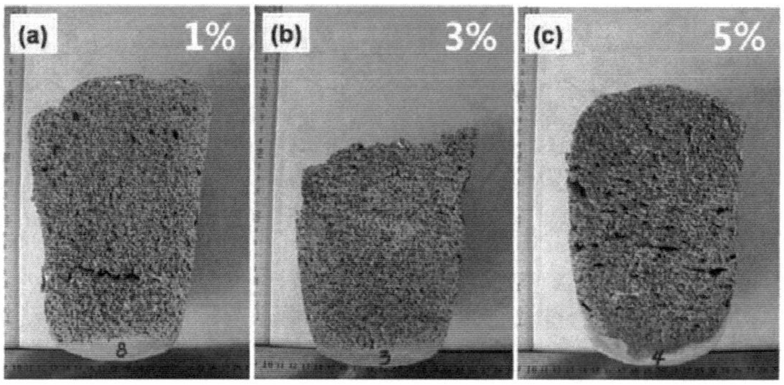

Figure 4. Macrostructures of foamed aluminum produced by different additions of shell at 670 °C.

(a) 1 %; (b) 3%; (c) 5%

The cut sections of the foams were examined for number of PPI. Pores per inch were measured along vertical (foaming direction) and horizontal direction of cut section. Then the average number of PPI was calculated and the density of aluminium foam, relative density and porosity was calculated. The variation of PPI and relative density with shell addition are depicted in figures 5 and 6 respectively.

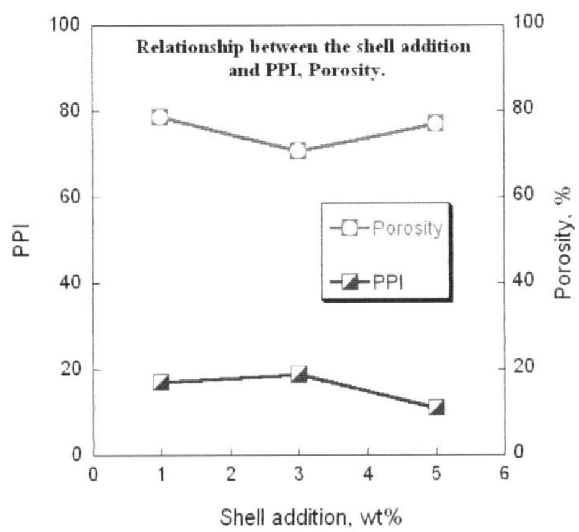

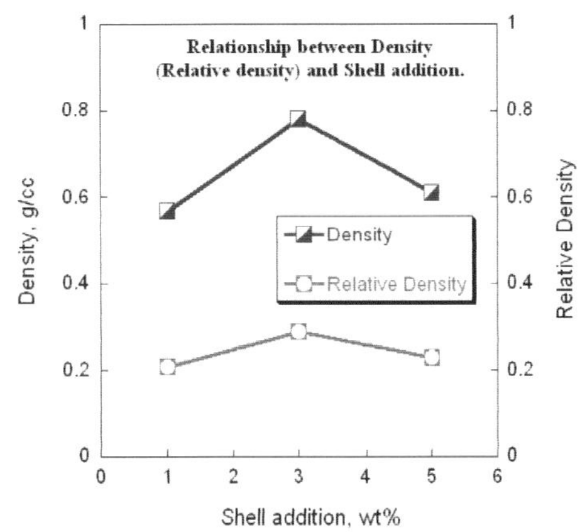

Figure 5(left). Relationship between the PPI (Porosity) of foamed aluminum and shell addition.

Figure 6(right). Relationship between the relative density (density) of foamed aluminum and shell addition.

Pores per inch, density, relative density and porosity of the foam produced with different addition of shell are tabulated in table 1.

Table 1. Shell addition and other physical properties.

Shell addition (mass %)	Pore per inch (PPI)	Density (g/cc)	Relative density	Porosity (%)
1	17	0.57	0.21	78.73
3	19	0.78	0.29	70.89
5	11	0.61	0.23	77.24

A SEM micrograph of the fracture structure of foamed aluminum with 5% by weight of shell is shown in figure 7. The foam has remarkably thin cell walls (25 μm), which therefore leads to a film structure which is easily broken. Eventually, the cells will vanish. These thin cell walls are caused by the excessive foaming. These experimental results indicate that the higher the shell amount, the more gaseous carbon dioxide is released which can increase the foam porosity. However, excess released gas escaping through the melt results in low foaming efficiency. Therefore, the appropriate addition amount of shell is 3 % by weight for production of a uniform cell structure from the viewpoints of efficiency and economy [11, 12].

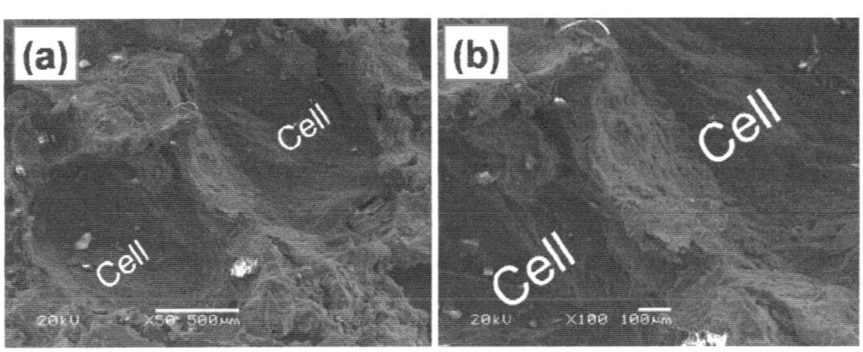

Figure 7. SEM image of fractured surface of foamed AlSi7Mg alloy.

(a) Cell image; (b) Cell image

4. Conclusions

The factors which affect the foaming in a foamed aluminum casting process were investigated. Properly controlling the shell content of the melt leads to the production of foamed aluminum which contains a uniform cell structure of high porosity. An optimum foamed aluminum with air bubbles of 1.3 to 3 mm diameter and a uniform distribution with 71% porosity is obtained by the addition of 3% shell at 670°C.

References

[1] Gibson L J and Ashby M F 1999 *Cellular Solids, Structure and Properties*, 2nd Edition, Cambridge University Press, Cambridge.

[2] Ashby A G, Evans N A, Fleck L J, Gibson J W, Hutchinson M F and Wadley H N G 2000 *Metal Foams: A Design Guide*, Butterworth Heinemann, Oxford.

[3] Soloki A and Esmailian M 2015 *Metal Mater. Trans. B* Carbonate-Foaming agents in aluminum foams: Advantages and Perspectives **46** 1052-57.

[4] Golestanipour M, Mashhadi H A, Abravi M S, Malekjafarian M M and Sadeghian M H 2011 *Mater. Sci. Technol.* Manufacturing of Al/SiC_p composite foams using calcium carbonate as foaming agent, **27** 923-7.

[5] Haesche M, Lehmhus D, Weise J, Wichmann M, Cristina I and Mocellin M 2010 *J. Mater. Sci. Technol.* Carbonates as foaming agent in chip-based aluminium foam precursor, **26** 845-50.

[6] Byakova A V, Gnyloskurenko S V, Sirko A I, Milman Y V and Nakamura T, 2006 *Mater. Trans.* The role of foaming agent in structure and mechanical performance of Al based foams, **47** 2131-36.

[7] Nakamura1 T, Gnyloskurenko S V, Sakamoto K, Byakova A V and Ishikawa R 2002 *Mater. Trans.* Development of new foaming agent for metal foam, **43** 1191-96.

[8] Nadella R, Sahu S N and Gokhale A A 2010 *Mater. Sci. Technol.* Foaming characteristics of Al-Si-Mg (LM25) alloy prepared by liquid metal processing, **26** 908-13.

[9] Yang C C and Nakae H 2000 J. Alloys Compd. Foaming characteristics control during production of aluminum alloy foam, **313** 188-91.

[10] Vinod Kumar G S, Chakraborty M, Garcia-Moreno F and Banhart J 2011 *Metal Mater Trans A.* Foamability of $MgAl_2O_4$ (Spinel)-Reinforced aluminum alloy composites, **42** 2898-2908.

[11] Praveen Kumar T N, Venkateswaran S and Seetharamu S 2015 *Trans. Indian Inst. Met.* Effect of grain size of calcium carbonate foaming agent on physical properties of eutectic Al-Si alloy closed cell foam, **68** S109-S12.

[12] Katayose A, Yokose R and Makuta 2016 *Mater. Lett.* T Fabrication of closed-cell porous metals by using ultrasonically generated microbubbles, **185** 211-13.

Acknowledgments

The authors are gratefully acknowledging Dalian Jiao-Tong University (Project No. 779031) and the NSF of Liaoning Province (Project No. 20170540140), for the financial support for this Project.

IOP Conf. Series: Materials Science and Engineering **649** (2019) 012003 doi:10.1088/1757-899X/649/1/012003

The Simulation of Magneto-Mechanical Properties of Magnetorheological Elastomers

Dan Zhao[1, a*], Bing Wang[2, b], Zhenghang Zhao[3, c], Shuaihu Wang[1, d], Song Yang[1, e] and Na Dong[1, f]

[1]College of Mechanical and Electrical Engineering, Harbin Engineering University, China
[2]Shanghai Marine Equipment Research Institute, CSIC, Shanghai, China
[3]School of Mechanical Engineering, Heilongjiang University of Science and Technology, China

zhaodan@hrbeu.edu.cn, wangbinghaiyuebox@163.com, 215484869@qq.com, Wangshuaihu@hrbeu.edu.cn, Yangsong@hrbeu.edu.cn, Dongna@hrbeu.edu.cn

Abstract. Magnetorheological elastomers is an intelligent material which can sense and judge the external magnetic field and change its mechanical properties. In this paper, functional carbon nanotubes are added to polyurethane magnetorheological elastomers. According to the micro-mechanical finite element calculation, the mechanical parameters of the equivalent matrix composed of carbon nanotubes and polyurethane elastomers are obtained. Multi-scale simulation analysis of the magneto-mechanical properties of isotropic magnetorheological elastomers with different volume fractions of carbon nanotubes in tensile, compressive and shear modes using magnetic coupling software. The simulation results show that the magnetorheological elastomer with larger volume fraction of carbon nanotubes has higher storage modulus in the three modes than the magnetorheological elastomer with smaller volume fraction of carbon nanotubes.

1. Introduction

The research of mechanical analysis of magnetorheological elastomer (MRE), is the coupling research of the magnetic-particle under the magnetic field force and the Shear or tension force. In recent years, many scholars have studied on the magneto-mechanical properties of MRE. In 1999, Ginder [1] used dipole model and finite element method to calculate and analyze the shear stress-strain relationship of MRE under different magnetic fields. In 2004, Y. Shen [2], according to the characters of the MRE particle chains and the nonlinearity of matrix materials, established a model for the magneto-mechanical modulus of MRE. In 2007, Kallio [3] studied on the mechanical properties of MRE in compression mode, the results showed that the compression modulus of anisotropic MRE formed by magnetic particles is greater than that of isotropic MRE. Zhu [4], calculated the magnetic field distribution, shear modulus, relative permeability and other parameters in magnetorheological elastomers by finite element method. Sun [5, 6] carried out numerical simulation on the mechanical properties of magnetorheological elastomers with or without external magnetic field through RVE based on periodic boundary conditions, then compared and analyzed the result with the theory. Zhao [7], based on the Meso-Mechanics RVE, studied on the magneto-mechanical of mechanical properties and magneto-mechanical shearing of multi-chain MRE by theoretical analysis and numerical simulation, the results showed that the initial compressive stress of MRE decreases with the application of magnetic field, and the shear modulus increases with the application of magnetic field.

In this paper, carbon nanotubes were added into the magnetorheological elastomer of polyurethane. Since the geometrical size of carbon nanotubes was nanoscale and carbonyl iron powder (CIP) was micron scale, meso-mechanics and finite element analysis theory were used to conduct multi-scale simulation analysis on the mechanical properties of MRE to study the impact of carbon nanotubes on the mechanical properties of MRE.

2. Analysis of Mechanical Properties of MWCNTs/ Polyurethane Elastomer Equivalent Matrix

In this paper, the process of the multi-walled carbon nanotube model constructed by figure 1. The length of MWCNTs $L_{MWCNT} = 12\mu m$, The outside diameter of MWCNTs is 10nm,inner diameter is 4nm.MWCNTs and the polyurethane elastomer are isotropic materials, The performance parameters of the material are $E_{MWCNT} = 1TPa, v_{MWCNT} = 0.3, E_{PU} = 39.4MPa, v_{PU} = 0.42$.

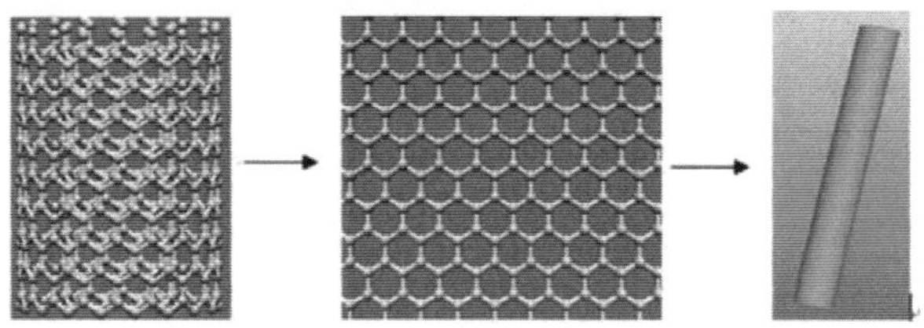

(a) MWCNTs molecular model (b) frame structure model (c) Cylindrical continuum model

Figure 1. The evolution process of MWCNTs.

In recent years, studies have found that MWCNTs can change the mechanical properties of PU. With the increase of the dosage of MWCNTs, the stiffness, strength and elasticity of polyurethane elastomer (PU) increase. The relevant experimental results also showed that, when MWCNTs are added to the composite material, agglomeration will occur when the volume fraction of MWCNTs is 1:1 to the specific gravity of the other filler, and all properties of the composite material will decline. According to the previous research results, carbonyl iron powder with volume fraction of 18% is selected in this paper to study the influence of carbon nanotubes with volume fraction of 4%, 8% and 12% on the mechanical properties of magnetorheological elastomers. First, establish the equivalent matrix two-dimensional representative volume unit (RVE) composed of MWCNTs/PU, and uses carbon nanotubes as the reinforcement of polyurethane matrix. The model size of RVE was selected as 160nm×160nm×160nm, and the number of particles of carbon nanotubes was calculated as 4, 6 and 10 respectively. The model is shown in figure 2.

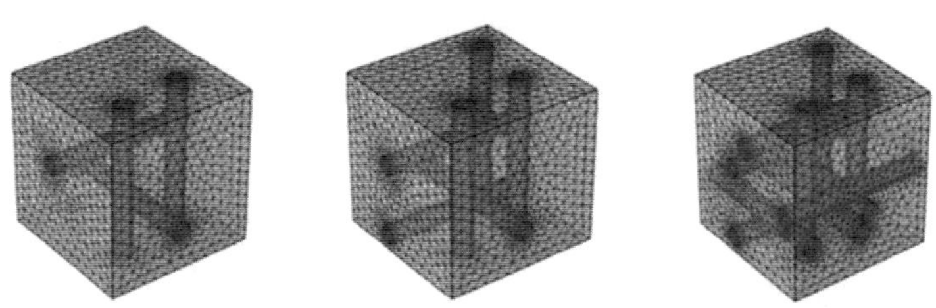

The volume fraction of carbon nanotube (a) 4% (b) 8% (c) 12%

Figure 2. Equivalent models of carbon nanotubes of different volumes.

The position vector of RVE is x, the volume is V, take a point, the stress is $\sigma(x)$,the strain is $\varepsilon(x)$, Thus, the average stress of the volume element is:

$$\bar{\sigma} = \frac{1}{V}\int_V \sigma(x)\, dV \tag{1}$$

The average strain is:

$$\bar{\sigma} = \frac{1}{V}\int_V \sigma(x)\,dV \tag{2}$$

Under small deformation, the effective stiffness matrix of the composite material is:

$$\bar{\sigma}_{ij} = C^*_{ij}\bar{\varepsilon}_{ij} \tag{3}$$

The $\bar{\sigma}_{ij}$ is the average stress of the composite material, $\bar{\varepsilon}_{ij}$ is the average strain of the composite material. According to the equation (1-3), the effective stiffness matrix or effective elastic modulus of RVE can be obtained.

$$\bar{\sigma} = \frac{1}{V}\sum_{i=1}^{i=n} \sigma_i v_i \tag{4}$$

$$\bar{\varepsilon} = \frac{1}{V}\sum_{i=1}^{i=n} \varepsilon_i\, v_i \tag{5}$$

V is the volume of RVE, $\sigma_i, \varepsilon_i, v_i$ is the corresponding RVE unit which called i unit, its stress, strain and volume.

C^*_{ij} is a 6 by 6 matrix, about the isotropic material, the unknown quantities are only C^*_{11} and C^*_{12}. The solution is to set a set of boundary load conditions, using volume averaging method to calculate the six average stress component and the six average strain, then according to the equations the effective stiffness matrix C^*_{ij}, of RVE, can be solved. The flexibility stiffness matrix S^*_{ij}, can be obtained by inverting the effective stiffness matrix, $S^*_{ij} = C^{*-1}_{ij}$. The effective mechanical elastic modulus of composite materials can be obtained.

$$\begin{vmatrix}\bar{\sigma}_{XX}\\ \bar{\sigma}_{YY}\\ \bar{\sigma}_{ZZ}\\ \bar{\tau}_{YZ}\\ \bar{\tau}_{XY}\\ \bar{\tau}_{XZ}\end{vmatrix} = \begin{bmatrix} C^*_{11} & C^*_{12} & C^*_{12} & 0 & 0 & 0 \\ C^*_{12} & C^*_{11} & C^*_{12} & 0 & 0 & 0 \\ C^*_{12} & C^*_{12} & C^*_{11} & 0 & 0 & 0 \\ 0 & 0 & 0 & (C^*_{11}-C^*_{12})/2 & 0 & 0 \\ 0 & 0 & 0 & 0 & (C^*_{11}-C^*_{12})/2 & 0 \\ 0 & 0 & 0 & 0 & 0 & (C^*_{11}-C^*_{12})/2 \end{bmatrix} \begin{vmatrix}\bar{\varepsilon}_{XX}\\ \bar{\varepsilon}_{YY}\\ \bar{\varepsilon}_{ZZ}\\ \bar{\gamma}_{YZ}\\ \bar{\gamma}_{XY}\\ \bar{\gamma}_{XZ}\end{vmatrix} \tag{6}$$

In this paper, carbon nanotubes are equivalent to cylinders and the volume fraction of carbon nanotubes is overestimated, but it does not have a great impact on the subsequent analysis.The results show that, adding carbon nanotubes to polyurethane elastomers, when the content of carbon nanotubes between 0% to 12%, as the volume fraction of carbon nanotubes increases, the young's modulus of matrix is observably increased. Therefor, appropriate carbon nanotubes can obviously enhance the mechanical properties of matrix materials. The Young's modulus and Poisson ratio has been shown in table 1.

Table 1. Young's modulus and Poisson ratio.

E_1	E_2	E_3	v_1	v_2	v_3
201MPa	368MPa	549MPa	0.37	0.36	0.35

IOP Conf. Series: Materials Science and Engineering **649** (2019) 012003 doi:10.1088/1757-899X/649/1/012003

3. Numerical Simulation of Polyurethane Magneto-mechanical Properties

The parameters of the magnetic - mechanical coupling numerical simulation model of polyurethane MRE with carbon nanotubes are shown in table 1 and table 2.

Table 2. Parameters of the numerical simulation model of polyurethane MRE with carbon nanotubes.

RVE	Square length	68um
Magnetic field	Magnetic field intensity	1T
Air	Relative permeability	1
Carbonyl iron powder	Young's modulus	210GPa
	Poisson's ratio	0.3
	Density	7900kg/m^3
	Relative permeability	5000
	Radius	8um

4. Magnetic Compression Simulation of Polyurethane MRE

Figure 3 shows the compressive deformation and strain diagram of MRE.

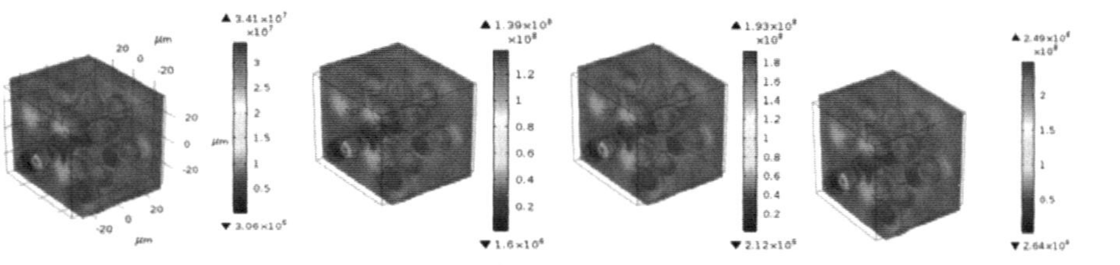

Figure 3. Compressive deformation and strain diagram of MRE.

Figure 3 is the stress cloud diagram of isotropic MRE with different volume fraction carbon nanotubes under magneto-compression. When added carbon nanotubes to ordinary MRE, the elastic modulus and stiffness of the equivalent matrix increase.

Figure 4 shows the stress-strain curves of isotropic MRE with different volume fractions of carbon nanotubes under magneto-compression. It is clearly shown in the figure that the initial stress curve of MRE decreases to a negative value after the magnetic field is applied, it shows that under the action of magnetic field, magnetorheological elastomer is subjected to the force opposite to the magnetic field. This shows that under the action of magnetic field, magnetorheological elastomer is subjected to the force opposite to the magnetic field.

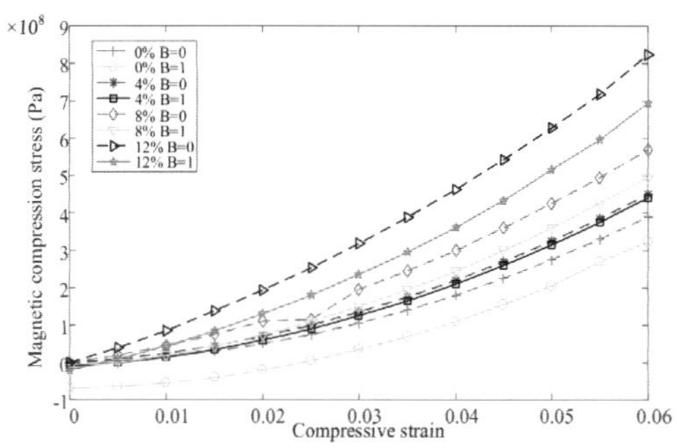

Figure 4. The stress-strain curves.

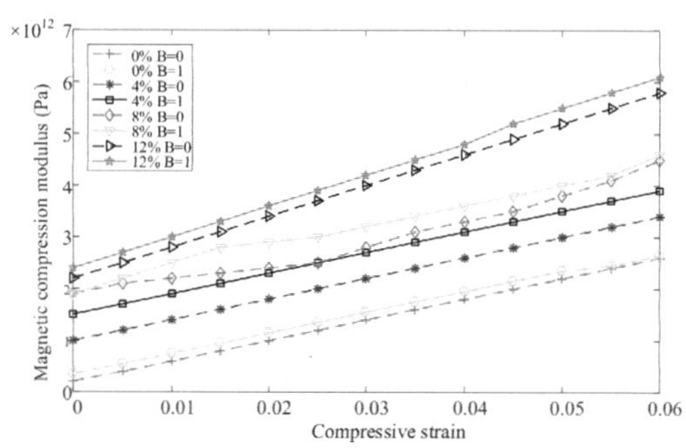

Figure 5. The module-strain curves.

Figure 5 is the magneto-strictive module-strain diagram of MRE of carbon nanotubes with different volume fractions under different field intensities drawn according to the magneto-strictive stress-strain curve according to figure 4. As shown in figure 5, under the same strain, the magneto-strictive modulus of MRE with field intensity is always greater than that without field intensity; Whether carbon nanotubes are added, the magneto-strictive modulus of MRE increases with the increase of compressive strain.

Magnetic shear simulation of polyurethane MRE

Figure 6 is the stress cloud diagram of isotropic MRE with different volume fraction carbon nanotubes under magneto-induced shear condition. First, it is shown that the comparison of the stress diagram that the external magnetic force and the external shear force are caused by the external magnetic field, it makes the surface stress of ferromagnetic particles obviously greater than that of elastomers, particularly, the surface stress of ferromagnetic particles in the same direction as the magnetic field is obviously greater than that of ferromagnetic particles perpendicular to the magnetic field. Even if the external magnetic field and external pressure are the same, however, the maximum stress value of MRE with a large volume fraction of carbon nanotubes added is larger than that of MRE with a small volume fraction of carbon nanotubes added.

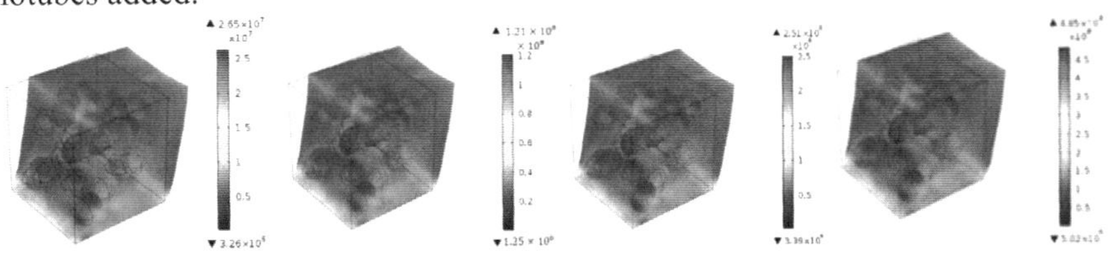

Figure 6. Shear deformation and stress map of MRE.

Figure 7 shows the simulation of magneto-strictive shear stress-strain curves of MRE with different volume fractions of carbon nanotubes. Firstly, it can be seen from the figure that the magneto-strictive stress increases with the increase of strain value, and the initial magneto-strictive stress increases with the application of external magnetic field. When the external magnetic field is applied, the magneto-strictive stress-strain curve starts from non-zero. Secondly, under the same shear stress condition, the strain value of magnetorheological elastomers with small volume fraction added carbon nanotubes was smaller.

After applying external magnetic field to MRE, the magneto-strictive modulus of MRE decreases, while the magneto-compression and shear modulus of MRE increase. Figure 8 shows the magneto-strictive module-strain curves of MRE with different volume fractions of carbon nanotubes.

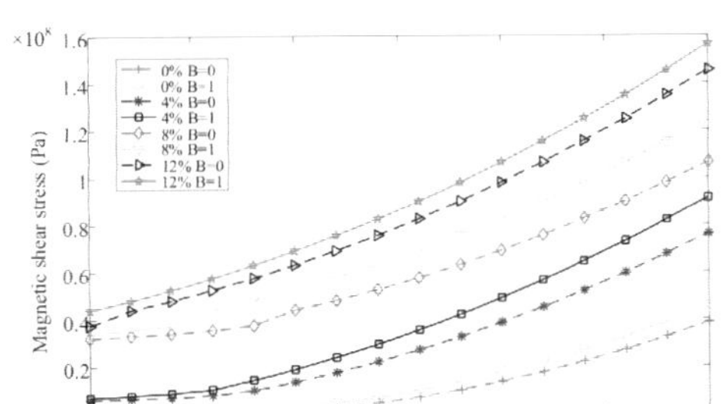

Figure 7. The shear stress-strain curves.

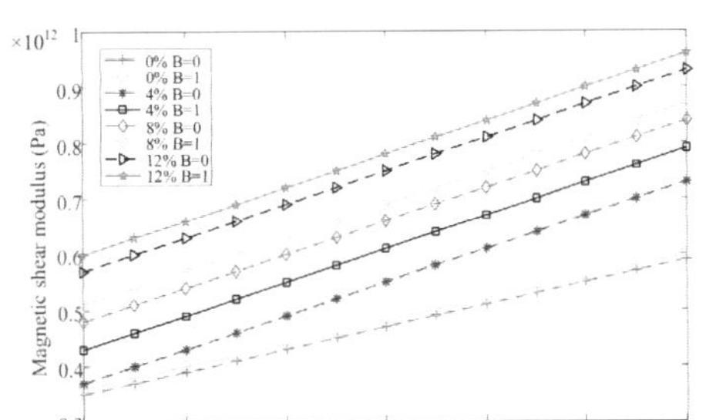

Figure 8. The shear module-strain curves.

Conclusion

This paper is based on Meso-Mechanics theory and finite element analysis software, simulated and analyzed the mechanical properties of MRE with different volume fractions of carbon nanotubes in the presence or absence of magnetic field at multiple scales. The results show that: In the compression state, the stress value of MRE decreases with the application of external magnetic field, magneto-strictive modulus increases; In the shear state, the stress value of MRE decreases with the application of external magnetic field, and the magneto-induced shear modulus increases; As the volume fraction of carbon nanotubes increases, the stress value and energy storage modulus of MRE increase, indicating that the addition of carbon nanotubes can improve the stiffness of MRE and optimize the mechanical properties of MRE.

References

[1] Ginder J M, Nichols M E, Elie L D, et al. 1999 Magnetorheological elastomers: properties and applications Proceedings of the 1999 Smart Structures and Materials on Smart Materials Technologies, Mar3-Mar4, Newport Beach, CA, USA, **3675** 131-8.
[2] Shen Y, Golnaraghi M F and Heppler G R 2004 *Intell. Mater. Syst. Struct.* Experimental Research and Modeling of Magnetorheological Elastomers **15** 27-35.
[3] Kallio M, Lindroos T, Aalto S, et al. 2007 *Smart Mater. Struct.* Dynamic compression testing of a tunable spring element consisting of a magnetorheological elastome **16** 506-14.
[4] Zhu 2007 *J. Comput. Mech.* Numerical analysis of some physical quantities of magnetorheological elastomers **24** 565-70.
[5] Sun 2015 *Func. Mater.* Prediction of mechanical behavior of magnetorheological elastomers based on RVE method **(4)** 4128-32.
[6] Sun 2015 Numerical simulation of finite deformation of magnetorheological fluid and study of this model [D]. Northwestern polytechnical university.
[7] Zhao 2016 *J. hunan univ. Eng. (self-edition)* Numerical simulation of magnetorheological properties of magnetorheological elastomers **26** 84-9.

Acknowledgments

This research is supported by the National Natural Science Foundation of China (No. 51675111, No. 51775123)

Modeling and Simulation Analysis of a Two-degree-of-Freedom Nonlinear Piezoelectric Energy Harvester

Junxiang Jiang[1], Shaogang Liu[1*], Junchao Wu[2] and Lifeng Feng[1]

[1]College of Mechanical and Electrical Engineering, Harbin Engineering University, Harbin, China
[2]Shanghai Marine Equipment Research Institute, CSIC, Shanghai, China

jiangjunxiang2014@163.com, liu_shaogang@hotmail.com, 13801815159@163.com, feng_lifeng@yeah.net

Abstract. In order to improve energy harvesting efficiency, this paper shows the design of a two-degree-of-freedom (2-DOF) nonlinear piezoelectric energy harvester. The proposed harvest is realized by introducing magnetic attraction to a 2-DOF piezoelectric cantilever beam in order to broaden the energy harvesting bandwidth. Electromechanical equations describing this nonlinear system are presented. The simulation results confirmed the advantage of this new 2-DOF nonlinear harvester. Performance comparison shows that the 2-DOF nonlinear harvester has much higher energy harvesting efficiency than a conventional one-degree-of-freedom (1-DOF) nonlinear one.

1. Introduction

Piezoelectric materials are the most common intelligent materials in electromechanical and vibration systems [1]. The previous studies have shown that the energy conversion efficiency of vibration energy harvester is the highest in resonance state [2]. When the operating frequency deviates from the natural frequency of the beam, the output power of energy harvesting structure decreased obviously. Different forms of energy harvesting structures for the broadband and multi-direction vibration excitation need to be studied. Many researches have been conducted on designing energy harvester such as tuning methods [3], frequency-up methods [4], nonlinear methods [5] and so on.

In the design of piezoelectric energy harvester, the nonlinear method is an effective method to improve the performance of the harvester. Erturk et al. introduced the bistable concept into the design of energy harvester to gain much higher voltage [6]. Zhou et al. proposed the theoretical model and experimental investigations of a broadband piezoelectric energy harvester with a magnetic field which generated higher energy output over a wider range of frequency [7]. Zhao et al. designed a new two degree of freedom bistable piezoelectric vibration energy to broaden the working frequency band of the piezoelectric energy harvester [8]. To improve energy harvester efficiency, a new 2-DOF nonlinear piezoelectric energy harvester is proposed in this paper.

In this study, the design principle and the dynamic model of the proposed 2-DOF nonlinear piezoelectric energy harvester is reported, and its dynamic performance is investigated by simulation.

2. Theoretical Modeling

The typical configuration of the 1-DOF nonlinear energy harvester and the position of the piezoelectric element are shown in figure.1(a). The nonlinear magnetic force created by magnet A and magnets (B, C) is more important to the nonlinear characteristics of harvester. Magnet A and magnets (B, C) are arranged by means of magnetic attraction. d_1 is the separation horizontal distance between the magnets (B, C) and the tip magnet A. d_2 is the gap distance between magnet B and magnet C. Monostable or bistable type of harvester could be obtained by adjusting the distances d_1 and d_2. Especially in the bistable state, the harvester exhibits the high-energy interwell motion. Higher voltage is generated and the effective bandwidth increases.

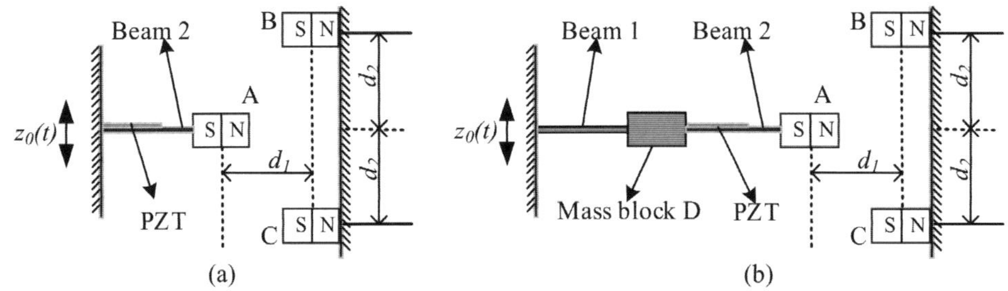

Figure 1. The structure diagram of the nonlinear piezoelectric energy harvester, (a) 1-DOF, (b) 2-DOF.

A 2-DOF cantilever nonlinear energy harvester is designed shown in figure1(b). Beam 1 and mass block D are added between the frame and magnet A. Then the harvester becomes a 2-DOF system and nonlinear phenomena can occur at the first two nature frequencies to broaden the effective frequency bandwidth. According to Newton's second law and Kirchhoff laws [9], the dynamic equation of the 2-DOF cantilever nonlinear system can be written as

$$\begin{cases} m_2(\ddot{z}_2+\ddot{z}_1+\ddot{z}_0)+c_2\dot{z}_2+k_2z_2+\alpha_{eq}V_L-F_z=0 \\ m_1(\ddot{z}_1+\ddot{z}_0)-c_2\dot{z}_2-k_2z_2+c_1\dot{z}_1+k_1z_1-\alpha_{eq}V_L=0 \\ \alpha_{eq}\dot{z}_2-V_L/R_L-C_p\dot{V}_L=0 \end{cases} \quad (1)$$

where $z_0=A\cos(\omega t)$ is the base excitation, z_1 is the relative displacement of the beam 1 to the base, z_2 is the relative displacement of the beam 2 to the beam 1. m_1, k_1, and c_1 is the equivalent mass, stiffness, damping of the beam 1. m_2, k_2, and c_2 is the equivalent mass, stiffness, damping of the beam 2. F_z is the attractive force. V_L is the voltage across the load resistance R_L, α_{eq} is backward electromechanical coupling in the dynamic equation, C_p is the clamped capacitance.

$$\begin{cases} \dot{x}_1=x_2 \\ \dot{x}_2=-\Omega_1^2x_1-2\zeta_1\Omega_1x_2+\mu\left(\Omega_2^2x_4+2\zeta_2\Omega_2x_5\right)+\dfrac{1}{m_1}\left(\alpha_{eq}x_3+F_z\right)-A\cos(\omega t) \\ \dot{x}_3=\dfrac{\alpha_{eq}}{C_p}x_5-\dfrac{1}{C_pR_L}x_3 \\ \dot{x}_4=x_5 \\ \dot{x}_5=\Omega_1^2x_1+2\zeta_1\Omega_1x_2-(\mu+1)\left(\Omega_2^2x_4+2\zeta_2\Omega_2x_5\right)-\dfrac{1}{m_2}\left((\mu+1)\alpha_{eq}x_3-F_z\right) \end{cases} \quad (2)$$

where $\mu=\dfrac{m_2}{m_1},\Omega_1=\sqrt{\dfrac{k_1}{m_1}},\Omega_2=\sqrt{\dfrac{k_2}{m_2}},\zeta_1=\dfrac{c_1}{2\sqrt{k_1m_1}},\zeta_2=\dfrac{c_2}{2\sqrt{k_2m_2}}.x_1=z_1$, $x_2=\dot{z}_1$, $x_3=V_L$, $x_4=z_2$ $x_5=\dot{z}_2$. Equation (2) can be solved by MATLAB software.

In the same way, we can obtain 1-DOF nonlinear piezoelectric energy harvester state equation.

3. Magnetic Analysis

The transverse displacement of magnet A is $z_2(t)$. The magnetic fields of dipoles B acting on dipole A is given by [10]

$$\mathbf{B}_{BA} = -\frac{\mu_0}{4\pi}\nabla\frac{\mathbf{m}_B \cdot \mathbf{r}_{BA}}{\left\|\mathbf{r}_{BA}\right\|_2^3} \tag{3}$$

where $\mathbf{r}_{BA}$ is the vector directed from the magnetic moment source of B to that of A. $\| \|_2$ denote Euclidean norm and Δ vector gradient operator. The flux density generated by the magnet C on magnet A can be evaluated by the similar formula. The potential energy of the total magnetic fields can be shown by

$$U_m = U_{mBA} + U_{mCA} = -\mathbf{B}_{BA} \cdot \mathbf{m}_A - \mathbf{B}_{CA} \cdot \mathbf{m}_A \tag{4}$$

where $\mathbf{m}$ is the magnetic moment vector.

The magnetic dipole moment vectors ($\mathbf{m}_A$, $\mathbf{m}_B$, $\mathbf{m}_C$) and the vectors ($\mathbf{r}_{AC}$, $\mathbf{r}_{BC}$) can be written as

$$\begin{cases} \mathbf{r}_{BA} = \begin{bmatrix} -d_1 & z - d_2 \end{bmatrix} \quad , \quad \mathbf{r}_{CA} = \begin{bmatrix} -d_1 & z + d_2 \end{bmatrix} \\ \mathbf{m}_C = \begin{bmatrix} M_C V_C & 0 \end{bmatrix} \quad , \quad \mathbf{m}_B = \begin{bmatrix} M_B V_B & 0 \end{bmatrix} \\ \mathbf{m}_A = \begin{bmatrix} M_A V_A \cos\theta & M_A V_A \sin\theta \end{bmatrix} \end{cases} \tag{5}$$

where M_A and V_A represent the magnetization intensity and the volume of magnet A. The meaning of the other parameters is the same. The subscript B, C stands for magnet B and magnet C.
Thus the magnetic force can be given by

$$F_z = \frac{\partial U_m}{\partial z} \tag{6}$$

The bistable state can be implemented by adjusting the separation distance d_1 and d_2.

4. Numerical Simulation

The parameters of the 2-DOF nonlinear energy harvester are shown in table 1. The magnetic potential can be obtained by equation (4). The separation distance d_1 is 7 mm and the gap distance d_2 vertically is 8 mm to make sure the harvester is a bistable system. Then the potential energy curve has two stable points and one unstable point.

Table 1. The parameters of proposed 2-DOF nonlinear piezoelectric energy harvester

Parameter	Value
Mass of beam 1 (m_1)	25 g
Mass of beam 2 (m_2)	4.8 g
Stiffness of beam 1 (k_1)	604 Nm^{-1}
Stiffness of beam 2(k_2)	133 Nm^{-1}
Damping of beam 1 (c_1)	0.0787 kgs^{-1}
Damping of beam 2 (c_2)	0.016 kgs^{-1}
Piezoelectric constant (d_{31})	285 C/N
Capacitance of the PZT (C_p)	25.7 nF
Magnetization vector (M_A, M_B, M_C)	1.2×10^6 Am^{-1}
Volume of magnet (V_A, V_B, V_C)	$100\times\pi$ m^3

IOP Conf. Series: Materials Science and Engineering **649** (2019) 012004 doi:10.1088/1757-899X/649/1/012004

The governing equation (2) and the 1-DOF nonlinear equation are solved by numerical method. The excitation amplitude is 5 m/s^2, the frequency ω is changing. The simulation parameters of the system are given in table 1. Figure 2 shows the root mean square RMS voltage responses of 2-DOF and 1-DOF nonlinear harvester at different excitation frequency. The system of 2-DOF has two effective voltage regions. The peak voltage is 1V at frequency 18Hz in the first region and the peak voltage is 0.6V in the second region.

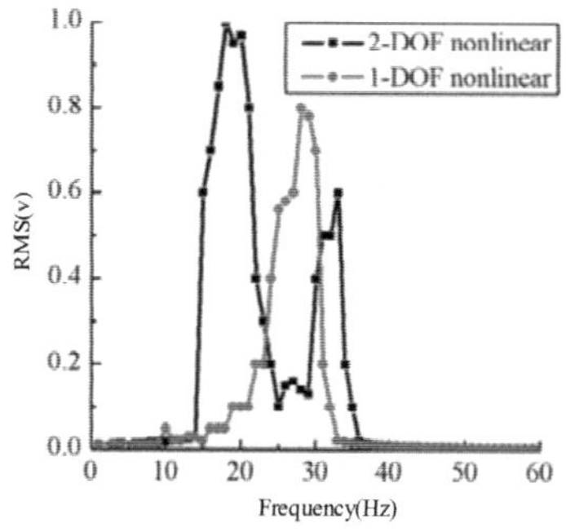

Figure 2. The theoretical voltage responses of 2-DOF and 1-DOF nonlinear harvester.

In the first region, at frequency 18Hz the voltage vs. velocity trajectory, voltage history and phase portrait are showed in figure 3(a), figure 3(b) and figure 3(c). The 2-DOF system shows very large amplitude electromechanical response and the large amplitude periodic orbit is observed. Figure 3(c) shows the system has periodic motions between two equilibrium points. So the voltage is larger. But the amplitude of motion decreases with increasing frequency. At frequency 24Hz the voltage vs. velocity trajectory, voltage history and phase portrait are showed in figure 3(d), figure 3(e) and figure 3(f). The system moves around an equilibrium point and the voltage is reduced to 0.2V. In the second region, compared with the first phase, the voltage has a similar trend. The voltage increases at frequency 30 Hz and with increasing frequency, then the RMS voltage decreases. The effective bandwidth of the first region is 5Hz and the effective bandwidth of the second region is 3Hz. so the total bandwidth is 8 Hz.

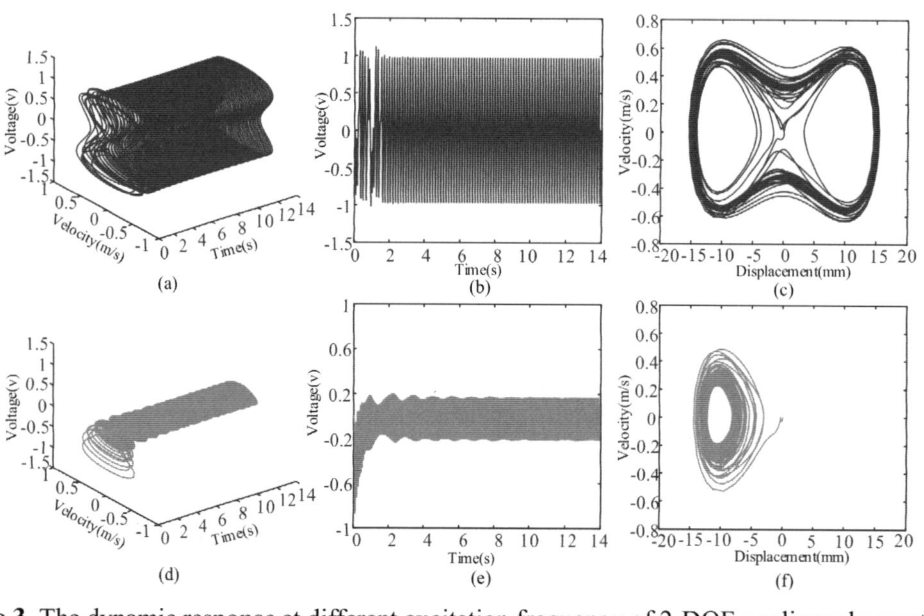

Figure 3. The dynamic response at different excitation frequency of 2-DOF nonlinear harvester for ω=18Hz (a) voltage vs. velocity trajectory (b) voltage history (c) phase portrait and ω=24Hz (d) voltage vs. velocity trajectory(e) voltage history (f) phase portrait.

But the 1-DOF nonlinear harvester only has one effective voltage region and the peak voltage is 0.8V. The effective bandwidth is 5Hz. The performance of 2-DOF nonlinear harvester is better than 1-DOF nonlinear harvester under the same conditions.

5. Conclusion

In this paper, a new 2-DOF nonlinear piezoelectric energy harvester with magnetic coupling is developed to increase the frequency bandwidth and improve the efficiency of energy harvesting from low frequency vibrations. The new harvester can be designed to harvest the vibration energy efficiently over a frequency range (16-21Hz) in the first region and effective frequency (30-33Hz) in the second region. The total effective bandwidth is about 1.6 times to that of the conventional 1-DOF nonlinear harvester. The simulation results indicate that compared 1-DOF nonlinear harvester the 2-DOF nonlinear harvester has a good dynamic characteristic.

References

[1] Harne R L and Wang K W, 2013 *Smart Mater. Struct.* A review of the recent research on vibration energy harvesting via bistable systems, **22** 023001.

[2] Ferrari M, Ferrari V, Guizzetti M, Andò B, Baglio S and Trigona C, 2010 *Sens. Actuators A Phys.* Improved energy harvesting from wideband vibrations by nonlinear piezoelectric converters, **162** 425-31.

[3] Hsu J C, Tseng C T and Chen Y S, 2014 *Smart Mater Struct.* Analysis and experiment of self-frequency-tuning piezoelectric energy harvesters for rotational motion, **23** 075013.

[4] Liu H, Lee C, Kobayashi T, Tay C J and Quan C, 2012 *Sens. Actuators A Phys.* Piezoelectric MEMS-based wideband energy harvesting systems using a frequency-up-conversion cantilever stopper, **186** 242-8.

[5] Cottone F, Vocca H and Gammaitoni L, 2012 *Phys. Rev. Lett.* Nonlinear energy harvesting, **102** 080601.

[6] Erturk A and Inman D J, 2011 *J. Sound Vib.* Broadband piezoelectric power generation on high-energy orbits of the bistable Duffing oscillator with electromechanical coupling, **330** 2339-53.

[7] Zhou S, Cao J, Inman D J, Lin J, Liu S and Wang Z, 2014 *Appl. Energy,* Broadband tristable energy harvester: Modeling and experiment verification, **133** 33-9.

[8] Dan Z, Minyao G, Chihang Z, Jundong W, Shaogang L and Tao W, 2018 *Mater. Res. Express*, Analysis of broadband characteristics of two degree of freedom bistable piezoelectric energy harvester, **5(8)**.

[9] Tang L and Yang Y, 2012 *J. Intell. Mater. Syst. Struct.* A multiple-degree-of-freedom piezoelectric energy harvesting model, **23** 1631-47.

[10] Lan C and Qin W, 2017 *Mech. Syst. Signal Process.* Enhancing ability of harvesting energy from random vibration by decreasing the potential barrier of bistable harvester, **85** 71-81.

Acknowledgments

This work is supported by the National Natural Science Foundation of China (No. 51675111).

Model Experiment of Construction Process of Long-Span Concrete Arch Bridge Rigid Skeleton

Shuangyang Zhang[1,*] and Ren-da Zhao[1]

[1]School of Civil Engineering, Southwest Jiaotong University, Chengdu 610031, China

syzhangswjtu@163.com, rendazhao@163.com

Abstract. High-speed railway passenger dedicated line have high demands for track smoothness, beipanjiang bridge as shanghai-kunming high-speed railway control project, mainspan 445m, constructed by using rigid skeleton, C80 high-strength concrete filled steel tube structure for the first time used in railway bridge; rigid skeleton in the process of construction has the traits: long span, long construction period, multi-step, mechanical behavior was extremely complex. In this paper, the design principle and similarity relation of rigid skeleton model are introduced, simplified the model construction process, and the finite element simulation analysis model is established, compared with measured values, it is shown that model test can well simulate the long span rigid skeleton arch bridge stress state in the construction process; the original construction steps have more safety reserves than simplified; the indicators of model test are within the scope of design, the results can provide technical support for the bridge construction.

1. Introduction

The stiff skeleton method, also known as embedded arch method, was first proposed by Czech engineer Joseph Melan in 1898, which greatly improved the spanning capacity of reinforced concrete arch bridge. The traditional stiff skeleton is mostly steel structure, and while increasing the stability of the skeleton, the material was wasted and poor economic. In the 1990s, China began to apply the concrete-filled steel tubular structure to the arch bridge. The truss structure composed of upper and lower chord of the concrete-filled steel tube structure greatly improved the overall stability, stiffenss and economy of the stiffened skeleton, and the concrete-filled steel tubular structure has the characteristics of high compressive strength and good ductility, so it's a very competitive construction method for long-span reinforced concrete arch bridges. China has built many long-span concrete bridges with concrete-filled steel tubular stiff skeleton method, such as Wanxian Yangtze River Highway Bridge with main span of 420m, Guangxi Yongning Yongjiang Bridge with main span of 312m, Yungui high speed railway Nanpanjiang Bridge with main span of 416m and Hukun high speed railway Bridge with main span of 445m, Japan also built the Qingye Bridge with a main span of 180m, Usagawa Bridge with a main span of 204m and the Takamatsu Bridge with a main span of 260m. Scholars of china and abroad have explored the ultimate span of reinforced concrete arch bridges[1, 2], the breakthrough is mainly in construction, and the application of the stiff skeleton method has promoted the further development of the span of concrete arch bridges[3-10].

2. Engineering Background

The total length of Beipanjiang River Bridge is 721.2m, and the mainspan is 445m high-strength concrete arch bridge and the height is 100m, the span ratio is 1/4.45, an arching coefficient of 1.6, and a single box of three chambers, contour and variable. Wide box section, the arch ring height is 9m, the arch ring outer wrapped material is C60 concrete, the steel tube concrete is used for the skeleton method construction, the steel tube is filled with C80 high-strength concrete. the highest design speed is 250km/h, and the reserves speed is 350km/h. The general layout of the bridge span is shown in figure 1.

Figure 1. Layout of Beipanjiang bridge.

3. Design and Production of the Stiff Skeleton Model

3.1. Similarity principle and design of stiff skeleton model

In principle, the geometrical dimensions of the model should be strictly according to the similarity ratio of 1/7.5. The span of the model is 59.333m, the height is 13.413m, the arch axis coefficient is 1.6, and the diameter of the steel tube is 102mm. The schematic diagram of the stiff skeleton structure is shown in figure 2 and figure 3.

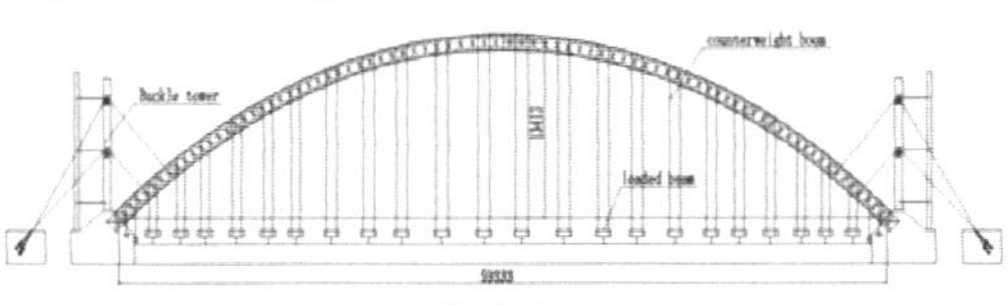

Figure 2. Elevation of rigid skeleton model.

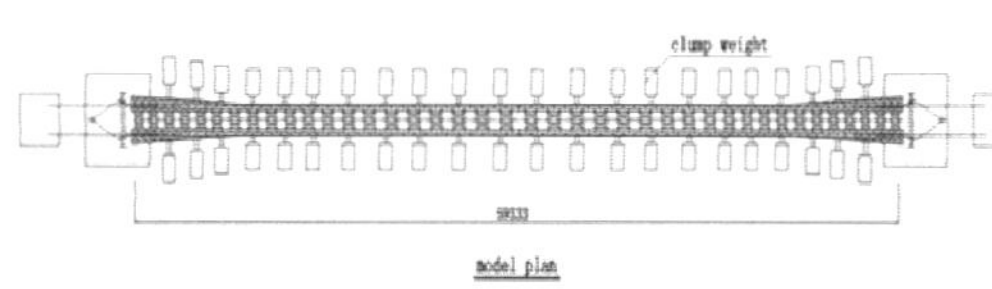

Figure 3. Plan of rigid skeleton model.

3.2. Model material and product

According to the model design principle and the purpose of the experimental research, the model is made of the same material as the actual structure, that is, C_E=1.0 and C_μ=1.0. In the process of model making, some model materials were properly adjusted due to factors such as material purchase, processing and on-site construction conditions. The comparison between the final stiff skeleton model and the main materials of the original bridge is shown in table 1.

Table 1. Material of rigid skeleton model and prototype.

structural element	Skeleton model	Prototype
Rigid frame steel tube	Q345B steel	Q370qC steel
Connection system Angle steel, joint plate, stiffening rib	Q235B steel	Q345B steel
Stube-filled concrete	CGM Bridge special high Strength self - compacting mortar	C80 concrete

IOP Conf. Series: Materials Science and Engineering **649** (2019) 012005 doi:10.1088/1757-899X/649/1/012005

According to the geometric similarity scale, the inner diameter of the stiff skeleton steel tube is only 94mm, and the concrete strength in the original bridge tube is C80, the coarse aggregate particle size is 5mm~20mm, and the fluidity is bad, so fine stone concrete was selected, and good test results were obtained at the same time as the construction. In order to prevent the tube blocking phenomenon during the pouring process, by collecting data and comparison, we determined to use the high-strength self-compacting CGM mortar for bridge. Since CGM mortar does not contain coarse aggregate, the material test results are shown in table 2.

Table 2. CGM bridge dedicated high-strength self-compacting mortar material test results.

Material	Cube compressive strength /MPa		elasticity modulus /MPa	Initial mobility /mm	30mins Fluidity retention value /mm
	3d	28d	28d		
CGM Mortar	43.6	66.6	3.52×104	373	263

4. Model Measured Points Arrangement

The test during the construction of the model includes the strain test and displacement test. The stiff skeleton model has a total of seven strain test sections, which are the two sides of the arch, L/8, L/4, L/2, 3L/4 and 7L/8 section, strain test sections shown in figure 4.

The strain test layout of each section of steel tube is shown in figure 5. G1~G8 are steel tube strain measuring points, and each test section is arranged with 8 steel tube strain measuring points. The whole bridge has a total of 56 steel tube strain measuring points.

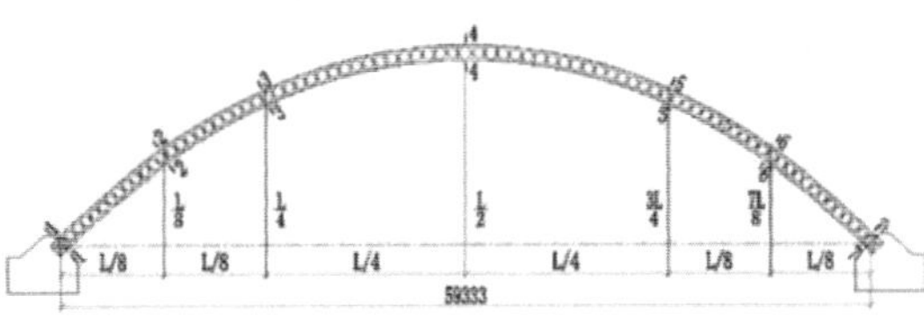

Figure 4. Layout of strain test section/mm.

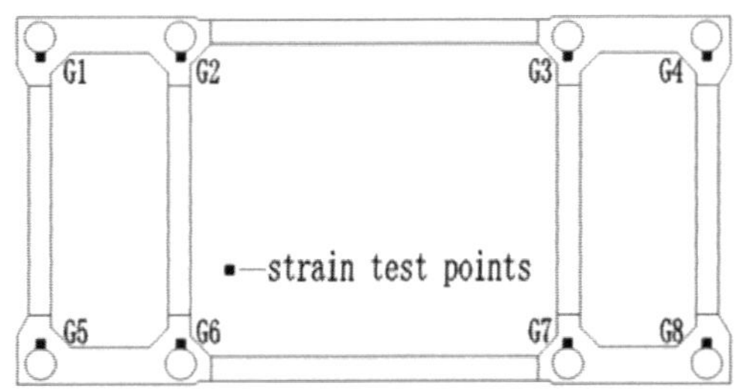

Figure 5. Layout of Strain measuring points.

A total of 7 displacement test sections are arranged in the stiff skeleton model, which are L/8, L/4, 3L/8, L/2, 5L/8, 3L/4 and 7L/8, as shown in figure 6.

A displacement measuring point is arranged on each of the upper outer steel tube of each displacement measuring point section, the displacement test and the strain test are simultaneously performed, and the displacement measuring points of each test section arrangement as shown in figure 7.

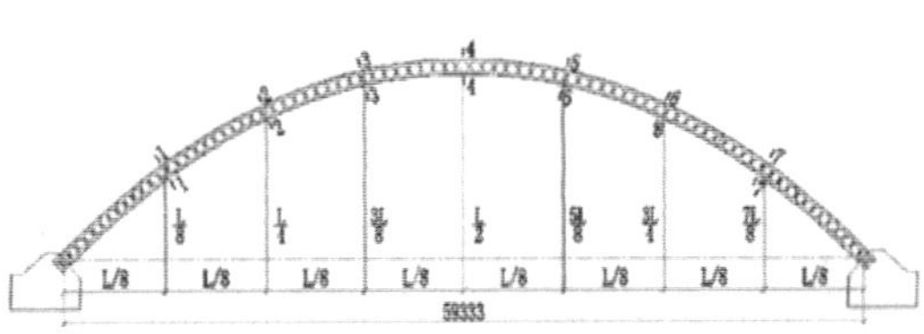

Figure 6. Layout of displacement test section/mm.

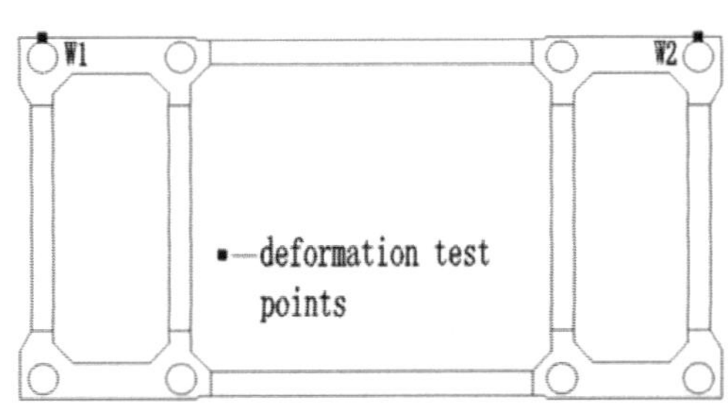

Figure 7. Layout of displacement measuring points.

5. Establish Finite Element Model and the Analysis of Test Results

5.1. Modeling ideas and model building

Establish a finite element calculation model with MIDAS/Civil to analysis The construction stage. The core concrete is subjected to less compressive stress during the construction stage, and the tight force of the steel tube has not yet been exerted. Therefore, the interaction between the steel tube and the core concrete, geometric nonlinearity and the influence of material nonlinearity are not considered. The finite element model was shown as figure 8.

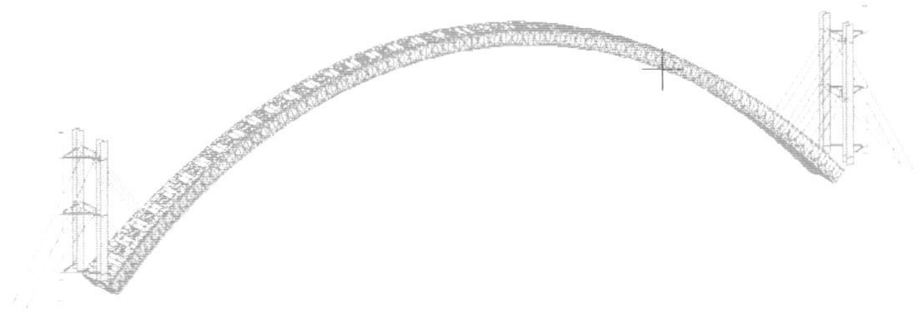

Figure 8. Finite element model.

5.2. Analysis of test results

The stiff skeleton model mainly controls the distribution of stress and displacement of the upper and lower string steel tubes under the main working condition as shown in figure 9-12. The changing curve of the main control sections steel tubes stress under various load conditions are shown in figure 9, 10. During the model test, the steel tube stress of the main control sections are in the 2# working condition to the 3# working condition increase is the largest, in which the stress of lower chord of the arch foot section and the cross-section of midspan increase 125.0 MPa, mainly because the concrete weight in the tube is loaded in the 3# working condition, and the applied load is large, but only the steel skeleton bear the load. In the process of web construction, in order to simplify the calculation, a one-time casting method is adopted, and the weight of the web is relatively large. Therefore, in the 15# working condition, the stress of the steel tube of the 2#~6# section increase greatly. Among them, the stress increase of the chord on the middle section is the largest, reaching 71.0 MPa, and the cross section of the arch foot section has formed a full section at this time, the concrete begins to be bear the load, and the stress increase of the steel tube is not so obviously.

The displacement changing curve of the main control sections under various load conditions are shown in figure 11, 12. The measured values of the 1/8 section is larger than the calculated values. In the 14th and 15th load cases, the measured values reached 1.5 times of the calculated values due to the second section of the full-section outer concrete is poured, the strength is not up to the design values, and the shrinkage and creep value of the early-age concrete is large, so that the measured value is significantly higher than the theoretical value, and the remaining sections are The measured values of the displacement agree well with the theoretical values, and the coincidence rate is about 90%.

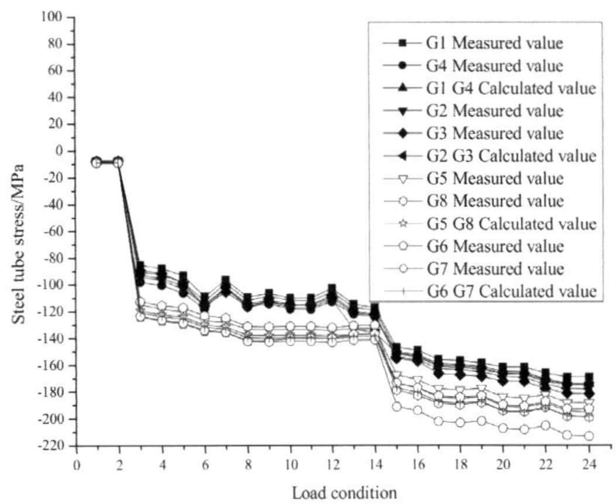

Figure 9. Steel tube stress measured values and calculated values of 1/8 section.

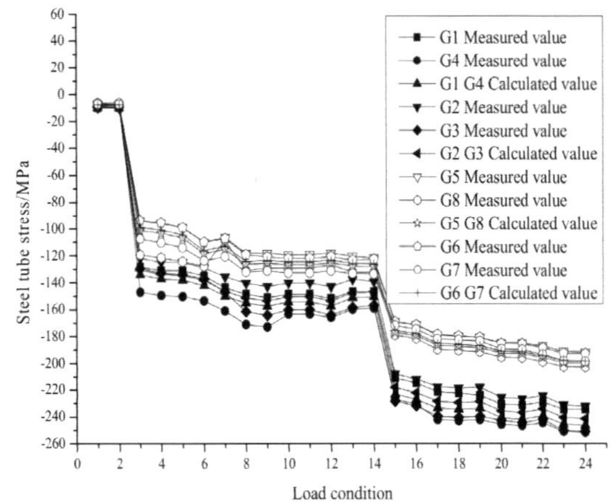

Figure 10. Steel tube stress measured values and calculated values of 1/2 section.

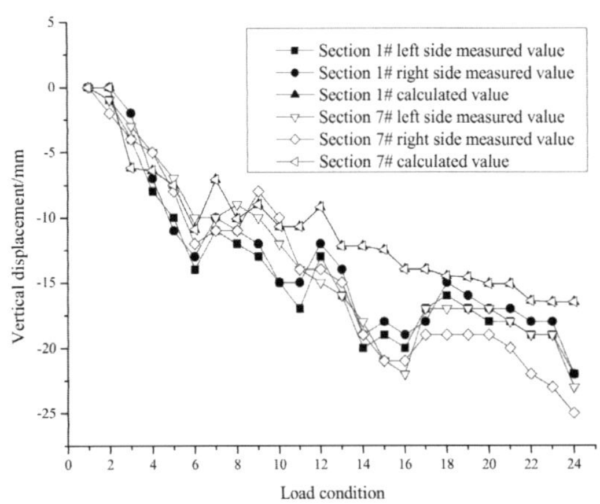

Figure 11. Displacement measured values and calculated values of 1/8 section

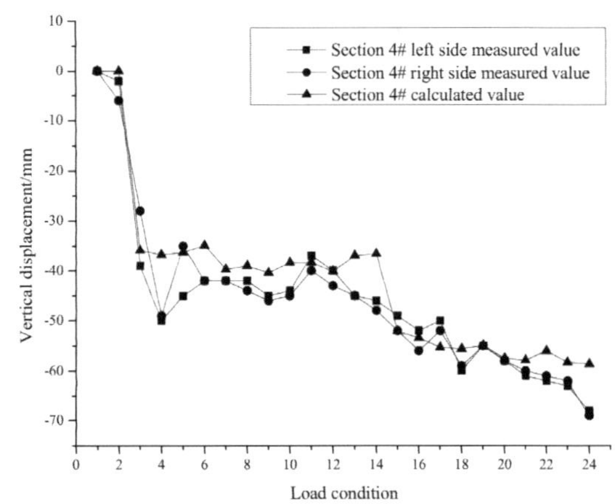

Figure 12. Displacement measured values and calculated values of 1/2 section.

6. Conclusions

Beipanjiang Bridge as the world's largest span high-speed railway stiff skeleton concrete arch bridge, the compression performance of high-strength CFST concrete structure can be fully exerted, and the structural stress and displacement changes are uniform and reasonable, which fully guarantees the safety and stability of the whole bridge construction.

The measured results and calculation results of the stiff skeleton model show that the main control sections have a large stress increase in the third load condition, and the stress increase of the chord and the mid-section of the arch section reaches 125.0MPa. and the 2#~6# sections steel tubes also appears large increase under the 15th load condition. The stress increase of the chord on the mid-section is 71.0MPa, In the construction process of the original bridge, it is necessary to pay attention to the stress changing of the steel tubes of the arch foot section and the mid-section to prevent the stress from exceeding the limit values and affect the structural safety.

During the construction of the model, due to the limitation of construction conditions, it is difficult to ensure the symmetric loading of the weights, so in the original bridge construction process should be ensure the symmetry of concrete pouring, to prevent excessive stress on the local steel tube, affect the stability of the full bridge structure.

References

[1] Xie H Q 2012 Study on structural type selection and mechanical behaviors of long-span railway concrete arch bridge with rigid skeleton [D]. Chengdu: Southwest Jiaotong University.

[2] Yang M H 2014 Model test study of concrete arch bridge with CFST stiffened skeleton [D]. Chengdu: Southwest Jiaotong University.

[3] Fan L A, Gong S L and Chen S T 2010 *J. Southwest Jiaotong Univ.* Model experiment of construction process of long-span steel truss bridge, **45(4)** 502-7.

[4] Hu Z J, Zhang M H, Kong X S, et al. 2014 *China J. Highway Transport* Static and dynamic experimental model for long-span steel-truss arch bridges, **27(9)** 82-9.

[5] Cui J, Sun B N and Lou W J 2004 *Eng. Mech.* Model test study on concrete filled steel-tube arch bridge, **21(5)** 83-6.

[6] Li Q, Tian X M and Zhang Q H 2003 *China Railway Sci.* A model test on long-span x-style arch bridge on railway, **24(1)** 88-92.

[7] Tian X M, Li Q and Zhang Q H *Brid. Const.* Study on static stiffness and model test of long-span x-style tied arch bridge in high-speed railway, **2002(2)** 12-5.

[8] Chacon R, Uribe N, Oller S 2013 *Experiment. Techniq.* Numerical validation of the incremental launching method of a steel bridge through a small-scale experimental study, **37** 1-14.

[9] Janke L, Czaderski C, Ruth J, et al. 2009 *Eng. Struct.* Experiments on the residual load-bearing capacity of prestressed confined concrete columns, **31(10)** 2247-56.

[10] Xiao L, Li X Z, Wei X, et al. 2013 *Eng. Mech.* Research and model test on fatigue performance of steel-concrete composite joint in space rigid frame, **30(5)** 83-9.

IOP Conf. Series: Materials Science and Engineering **649** (2019) 012006 doi:10.1088/1757-899X/649/1/012006

The Crack Nucleation Life of the TLP Flexjoint

Songyu Li[1, 2, 3], Peng Jia[3], Liquan Wang[3], Luyao Zhang[4], Kai Xu[3] and Feihong Yun[3]

[1]Shanghai branch, China Ship Science Research Center, Shanghai, China
[2]Shanghai Oriental Maritime Engineering Technology Co., Ltd, Shanghai, China
[3]College of Mechanical and Electrical Engineering, Harbin Engineering University, Harbin, China
[4]Shanghai Marine Diesel Engine Research Institute, Shanghai, China

lisongyu@702sh.com, jiapeng@hrbeu.edu.cn, wangliquan@hrbeu.edu.cn, 1097598039@qq.com, 525743293@qq.com, 664003732@qq.com

Abstract. Since the tension leg platform (TLP) needs to work in the ocean for a long time, it requires a high cost to replace any components. The working life is a key design element. The flexjoint is the crucial load-bearing component in the TLP mooring system. It has a multilayer metal/elastomer structure, and its working life generally depends on the rubber portion. In this paper, the fatigue life of flexjoint is calculated by crack nucleation method. Firstly, the dumbbell test piece is used to perform the uniaxial tensile fatigue test. The strain-fatigue life curve of the rubber material is obtained. The fatigue life of the flexjoint is calculated by the crack nucleation method. The fatigue test device of the flexjoint is developed, and the calculation result of crack nucleation life is verified by experiments.

1. Introduction

The tension leg platform is a deepwater production platform, which is mainly composed of a topside, a hull, and tendons and so on. The tendons are used for connecting TLP and the sea bottom. In order to meet the unique requirements of deepwater connection, a specific tendon connector is designed to complete the connection among the tendon, TLP and underwater pile foundation. In order to reduce the local bending stresses caused by vessel motions and environmental forces, a flexjoint is designed in each of the tendon connectors [1, 2]. It provides a certain rotational flexibility for the tendon connector, reduces the stress concentration at the end of the tendon, and also withstands a large axial pull. The flexjoint is manufactured by Laminated Elastomeric Component (LEC). LEC is a multilayer metal/elastomer structure in which the rubber is bonded to the metal. Compared to rubber materials, LECs can withstand higher compressive loads and increase compression stiffness while maintaining a relatively low shear stiffness. Based on the above advantages, LECs are used in many engineering equipments, such as the rubber bearing between the bridge and the pier, which can absorb the lateral vibration of the bridge [3]. A Leg Mating Unit (LMU) for offshore platform installation acts as a buffer during platform installation [4]. A solid booster flexible bearing joint in the solid rocket motor (SRM) allows the nozzle to deform in any given direction under engine control [5]. Spherical elastomeric bearings used in helicopter motors can withstand centrifugal and lateral forces to accommodate rotation, flip and leading lag movement [6]. Since TLP will work underwater for a long time and replacing any component requires a high cost, the design life becomes a key design element. The flexjoint is the crucial bearing component in TLP, so its fatigue performance is very significant. In practical work, the stress in the metal layer of flexjoint is small, so it's working life generally depends on the rubber part.

In general, rubber fatigue failure is mainly divided into two stages. In the first stage, the microscopic defects inside the rubber are gradually converted into visible cracks, ie, crack nucleation. In the second stage, The cracks slowly expand and eventually cause fatigue damage. According to this characteristic, two different methods are developed to calculate rubber fatigue life, ie, crack nucleation method and crack propagation method [7].

The crack nucleation method was first proposed by Wöhler [8] to calculate the fatigue life of the train axle. This method was then applied to rubber material by Cadwell et al [9]. The method considers that the fatigue life of rubber can be determined by the stress-strain data at each point on the test piece. This method often predicts the fatigue failure of rubber by strain, stress or energy. Since it can be obtained directly from the displacement data, strain is the most commonly used rubber fatigue prediction parameter. Suryatal et al. [10] used the maximum principal strain to predict the fatigue life of the compressed neoprene elastic pad. They found that increasing the maximum compressive load caused the maximum principal strain to rise, which in turn reduced the fatigue life. Li et al. [11] used the Mooney-Rivlin model to perform FE simulation on rubber specimens and used the maximum principal strain as the fatigue prediction parameter. The test results can be accurately fitted. Kim et al. [12] and Woo et al. [13] used Green Lagrangian strain as the fatigue prediction parameter of natural rubber dumbbell specimens. By comparing with the strain energy density, it was found that the Green Lagrangian strain would better fit test results. In this paper, the fatigue life of a flexjoint was calculated using the crack nucleation method. Firstly, the uniaxial tensile fatigue test was carried out using the dumbbell test piece, and the strain-fatigue life curve of the rubber material was obtained. The crack nucleation life of the flexjoint was calculated. A flexjoint fatigue test device was developed, and the calculation result was verified.

2. Rubber Crack Nucleation Life Calculation Method

Generally, due to processing defects or stress concentration, the rubber material has tiny defects inside. After under a long-term alternating loading, the micro-defects expand and eventually lead to crack nucleation. The fatigue life of a flexjoint calculated by the crack nucleation method is essentially the number of loading cycles required to produce an initial crack in the flexjoint. In mechanical design or fatigue analysis, the number of cycles of which a microscopic defect extends to a macrocrack of 1 mm length is defined as the crack nucleation lifetime. In order to calculate the crack nucleation life, two methods are generally used. One method uses nominal stress, and the other one uses strain. Since the rubber material is subjected to large deformation, the strain can be relatively more convenient to measure with displacement. Therefore, the nucleation life of the flexjoint is studied using the ε-N curve and equation. The ε-N curve of rubber is one of the intrinsic properties of rubber materials, which needs to be obtained through experiments.

The maximum principal strain is used as the fatigue life parameter here, because the initial crack of the rubber generally appears in a direction perpendicular to the maximum principal strain. The maximum principal strain in the flexjoint can be calculated by the FE method.

3. Rubber Uniaxial Tensile Fatigue Test

In order to obtain the ε-N curve, it is necessary to perform a uniaxial tensile fatigue test on the rubber. At present, the two most commonly used test pieces for this test are sheet-shaped dumbbell test pieces and column-shaped dumbbell test pieces, as shown in figure 1. The advantage of the cylindrical dumbbell test piece is that the uniaxial tension and uniaxial compression fatigue test can be performed simultaneously, and the fatigue life of the rubber material can be accurately predicted when the fatigue property of the rubber material is studied using the maximum principal strain. The advantage of the sheet-shaped dumbbell test piece is that it is easy to process. One only need to cut the rubber piece by the cutter. The test period for achieving fatigue life is shorter, and the fatigue life error predicted by using the test piece is not more than 28%, in engineering application. It can be used to replace the column dumbbell test piece.

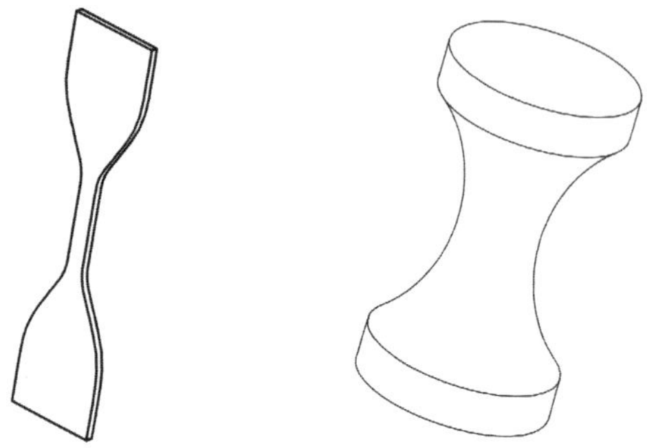

(a) Sheet-shaped (b) Column-shaped

Figure 1. Dumbbell type specimen of uniaxial tensile test.

Taking into account the processing cost and test cost, the sheet-shaped dumbbell test piece is selected to get the ε-N curve of the rubber material. The nitrile rubber is used, and the dumbbell test piece is provided by Tieling Yingang Rubber Factory. The total length of the test piece is 100 mm, the length of the stretched part is 20 mm, the width is 5 mm, and the thickness is 2 mm, as shown in Figure 2 (a). The test piece was designed to break quickly once a 1 mm length crack generated, so only the number of load cycles N_f at when the test piece broke was recorded.

The test was carried out at room temperature and a sinusoidal displacement fatigue load of 2 Hz was applied to the test piece using a Zwick 8801 tensile fatigue test rig, as shown in Figure 2(b). In this test, three different loading conditions were proposed. The minimum displacement was 0 and the maximum displacements were 15 mm, 20 mm and 25 mm, respectively. The corresponding maximum nominal strain ε was 0.56, 0.69 and 0.81. Three same dumbbell test pieces were used for each set of loading to eliminate individual differences.

(a) Dumbbell specimen (b) Zwick 8801

Figure 2. Uniaxial tensile fatigue test.

Figure 3 shows the ε-N curve obtained in experiment, whose horizontal and vertical coordinates are in a logarithmic coordinate system. According to the relationship between damage parameters and life, the ε-N equation is obtained by curve fitting as

$$N_f = 2.907 \times 10^5 \varepsilon_L^{-3.245} \quad (1)$$

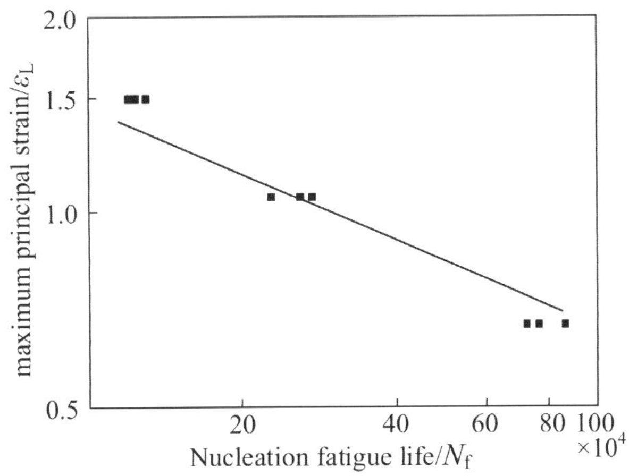

Figure 3. ε-N curve of NBR.

4. Nucleation Life of Flexjoint

It is assumed that the failure of the flexjoint does not occur in the metal portion. According to the above, the crack nucleation life of the rubber material under corresponding loading condition has been obtained by experiments. At this time, the FE method is needed to calculate the maximum principal strain of the rubber part in the flexjoint. In this section, the nitrile rubber flexjoint with three layers of metal plates, as shown in figure 4, was used. The crack nucleation life of the sample subjected to compression and rotational loads at the same time was calculated.

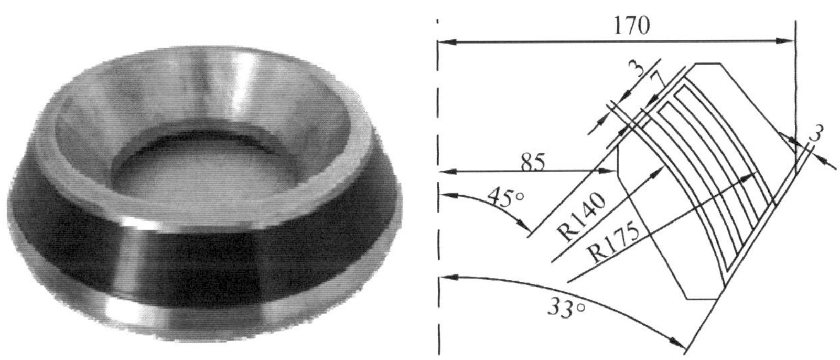

Figure 4. Flexjoint prototype.

Firstly, the FE method is used to simulate the whole flexjoint, and the 1/2 three-dimensional FE model of the rubber layer is established. The rubber was simulated using the Yeoh constitutive model. The parameters of the nitrile rubber material, as shown in table 1, was obtained by fitting the data of uniaxial tensile, simple shearing, biaxial tensile and volume compression test. A reference point RP-1 is established at the origin of the flexjoint. The top of the rubber layer is fixed, and the bottom is coupled to RP-1. In actual work, the rotational load of flexjoint comes from the horizontal movement of the tension leg platform, and the vertical compression load comes from the heave motion of the tension leg platform, which are both the displacement-controlled. Therefore, two analysis steps are set. In the analysis step-1, an angular displacement of 5° in the counterclockwise direction around the z-axis is applied to the RP-1. The displacement of 1.5 mm in the y direction is applied to the RP-1 and the deflection angle is kept constant in the analysis step-2. Figure 5 shows the maximum principal strain cloud diagram of the flexjoint. It can be seen that the maximum principal strain occurs in the first rubber layer of the flexjoint, and the maximum engineering principal strain value at the critical position is 1.894. The value is brought to the Eq.1, and the crack nucleation lifetime is obtained to be 37,000 times.

Table 1. Constitutive model parameters of NBR.

Yeoh	C_{10}[MPa]	C_{20}[MPa]	C_{30}[MPa]	D_1 [MPa^{-1}]	D_2[MPa^{-1}]	D_3[MPa^{-1}]
Parameters	0.307	-0.0109	0.000476	0.00168	1.29e-5	-1.45e-7

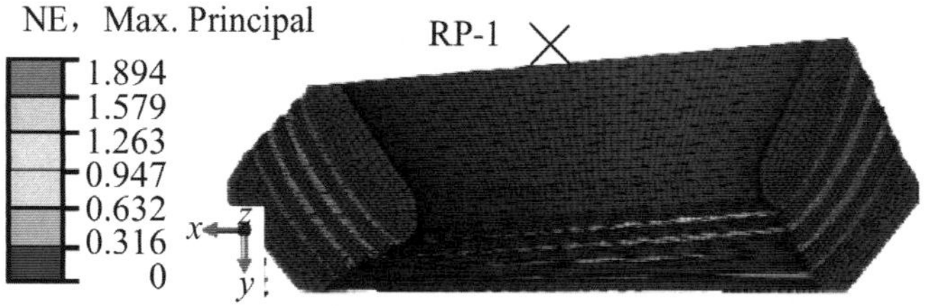

Figure 5. Maximum nominal principal strain of flexjoint under 0.5 ° angle and 100 kN compression.

5. Fatigue Test of Flexjoint

According to the foregoing description, the flexjoint is mainly subjected to a compressive load and a rotational load during operation. The rotational load comes from the swaying motion of TLP, and the compression displacement comes from the heave motion caused by the sway motion. Based on the same loading principle, this paper designs a flexjoint fatigue teser, as shown in figure 6. A hydraulic cylinder horizontally pushes the bottom of the swinging shaft, and the flexjoint will rotat by the swinging shaft. At the same time, the swinging shaft moves downward to compress the flexjoint. Since the fatigue tester needs to provide horizontal tension and thrust with opposite magnitudes in the same direction, and the load is large. By considering the control difficulty and construction cost, it is finally decided to use a hydraulic cylinder to provide loading. A solenoid valve is used to control the direction of the load. The limit switch is used to control the maximum horizontal displacement, and record the number of changes of the limit switch signal, which is recorded as the number of load cycles.

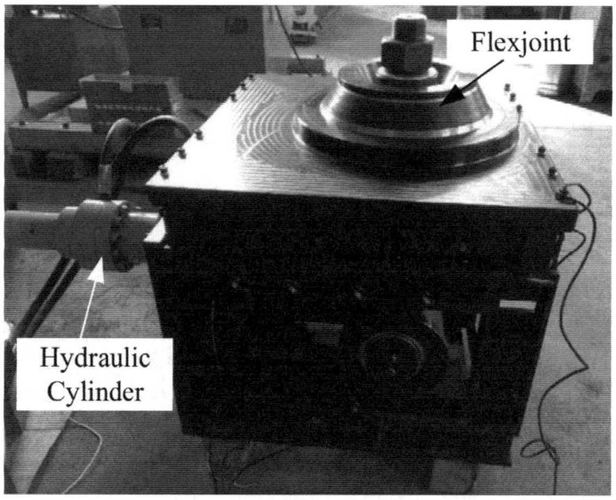

Figure 6. Fatigue tester of flexjoint.

The flexjoint fatigue test was carried out at room temperature, and the nitrile rubber flexjoint sample was the same one described in the previous section. According to the above description, the flexjoint load is displacement control with a rotational load of ±5 ° and a compressive load of 0 mm to 1.6 mm. Considering the transmission capacity of the hydraulic system, the load frequency was set to 0.5 Hz. When a significant crack appears on the flexjoint, the number of loadings at this time is the crack nucleation life. The nucleation fatigue life of the flexjoint obtained by the experiment is about 28,000 times, and the error with the theoretical calculation result is 32%, which is considered to meet the engineering requirements.

6. Summary

In this paper, the fatigue life calculation method of flexjoint is studied. The nucleation life of the flexjoint was calculated to be 37,000 times with the maximum engineering principal strain as the fatigue parameter. A flexjoint fatigue test device was developed. The nucleation fatigue life of the flexjoint crack obtained by the test was about 28,000 times, and the error with the theoretical calculation result was 32%, which was considered to meet the engineering requirements.

References

[1] Wald G M and Owens H S 1987 *Offshore Technol. Conf.* An Integrated Approach to TLP Tendon System Component Design.
[2] Kumar A and Hogan M E 2002 *Pipeline Technol.* Performance Related Parameters of Flexible Connections Used in Offshore Pipelines OMAE2002-28159 107-15.
[3] Rahman Bhuiyan A and Alam M S 2013 *J. Eng. Struct.* Seismic performance assessment of highway bridges equipped with superelastic shape memory alloy-based laminated rubber isolation bearing, **49** 396-407.
[4] Hu Z, Li X and Zhao W 2017 *J. Appl. Ocean Res.* Nonlinear dynamics and impact load in float-over installation, **65** 60-78.
[5] Zhang X, Liu Y and Ren J 2012 *Struct. Dynamics Mater. Conf.* Nonlinear finite element analysis of the SRM flexible joint. 53rd AIAA/ASME/ASCE/AHS/ASC Structures.
[6] Chen G S and Yang Y 2017 *J. Hangkong Dongli Xuebao/j. Aerospace Power* Stiffness design, simulation and test of laminated spherical elastomeric bearing, **30** 1512-19.
[7] Mars W V and Fatemi A. A 2002 *Int. J. Fatigue* literature survey on fatigue analysis approaches for rubber, **24** 949-61.
[8] Tee Y L and Loo M S 2018 *Int. J. Fatigue* Andriyana A. Recent advances on fatigue of rubber after the literature survey by Mars and Fatemi in 2002 and 2004, **110** 115-29.
[9] Cadwell S M, Merrill R A, Sloman C M, et al. 1940 *Ind. Eng. Chem. Anal. Ed.* Dynamic fatigue life of rubber, **12(1)** 19-23.
[10] Suryatal B, Phakatkar H, Rajkumar K, et al. 2015 *J. Surface Engineered Mater. Adv. Technol.* Fatigue Life Estimation of an Elastomeric Pad by ε-N Curve and FEA, **5** 85-92.
[11] Li Q, Zhao J-C and Zhao B 2009 *J. Eng. Failure Anal.* Fatigue life prediction of a rubber mount based on test of material properties and finite element analysis, **16** 2304-10.
[12] Kim W B, Lee H J, Kim J Y, et al. 2004 *Int. J. Fatigue* Fatigue life estimation of an engine rubber moun, **26** 553-60.
[13] Woo C S, Kim W D, Lee S H, et al. 2009 *J. Procedia Eng.* Fatigue life prediction of vulcanized natural rubber subjected to heat-aging, **1** 9-12.

Acknowledgments

The work described in this paper was fully supported by a grant from the Basic scientific reserach business fee in center colleges and universities (No. 3072019CF0705).

Study on the Damping Problem in Circular Plates of Radial Phonon Crystals

Di Mu[1], Haisheng Shu[1] and Yuqiang Su[1]

[1]Harbin Engineering University, College of Mechanical and Electrical Engineering, 150001, Harbin, China

shujs@hrbeu.edu.cn

Abstract. In order to perfect the theoretical framework of the radial phonon crystal circular plate, a damping type radial circular plate is constructed. The localization factor of damping is solved by transfer matrix method, and the finite element simulation and experimental verification are carried out. Furthermore, the effect of elastic modulus on the bandgap is investigated from the perspective of its frequency nonlinearity. The results show that the damping effect of viscoelastic materials has a significant effect on the bandgap, which is mainly reflected in the fact that the cut-off frequency of bandgap moves to the high frequency with the increase of material loss factor, thus broadening the bandgap. The damping effect of damping alloy has no obvious influence on the bandgap characteristic, which is related to the inherent impedance characteristic of the alloy.

1. Introduction

Phonon crystal, as a kind of periodic composite structure, plays an irreplaceable role in the field of vibration and noise reduction because of its unique band gap vibration characteristics[1-11]. In recent years, some researches have been made on the influence of the damping effect of phononic crystals on bandgap. In 2017, Chunyang Du et al. [12] designed a class of X-shaped locally resonant phononic crystal beams to solve the problem of over-damping, and pointed out the attenuation inhibition effect of such structures on the low-frequency band. In 2014, Siwen Zhang et al. [13] constructed a class of rod-shaped phonon crystals for viscoelastic damping, pointing out the regulating effect of loss modulus of viscoelastic materials on the energy band structure. Weikai Xu et al. [14] analyzed the bandgap characteristics of one-dimensional phonon crystals with damping. Frazier and Hussein [15] studied the overdamping phenomenon of viscoelastic materials. However, most of these studies focus on rod, beam and other structures, and do not involve the radial phonon crystal plate. In fact, our research group has conducted in-depth theoretical research on radial phononic crystal circular plates, including one-dimensional and two-dimensional structures, and made some research achievements [16-21]. In order to simplify the analysis, the damping problem of materials is generally ignored. In fact, damping is the inherent characteristic of all materials. For general metal materials, the material loss factor is very small, and the damping energy dissipation effect can be neglected in the research. However, for various damping materials widely used for vibration and noise reduction, such as damping alloy (Zn-Al alloy, Mn-Cu alloy) and viscoelastic material (epoxy resin, plastic), their values are large, respectively located at 0.05~0.2 and 0.1~5. As necessary, we must investigate the effect of damping energy dissipation.

On the basis of the above research, this paper further constructs a damped radial phonon crystal circular plate, deduces the corresponding transfer matrix, and analyzes the influence of damping on the bandgap characteristics of a radial circular plate from the perspective of local factor. The finite element method is used to analyze the frequency nonlinearity of elastic modulus. The results show that the damping energy dissipation characteristics of viscoelastic materials mainly affect the cut-off frequency of the bandgap and thus broaden the bandwidth.

2. Theoretical Analysis of Damped Radial Circular Plate

Figure 1 shows the model of 1D radial phononic crystal circular plate. Material 1 is damping material, while material 2 is the matrix material ignoring damping effect. E, ρ, v respectively represent young's modulus, density and Poisson's ratio of the two materials. As shown in the figure, a_1, a_2 are the radial width of damping material and matrix material, and r_0, h are the inner diameter of the innermost ring plate and the thickness of the thin circular plate. In this paper, the damping effect of material 1 on the bandgap is studied.

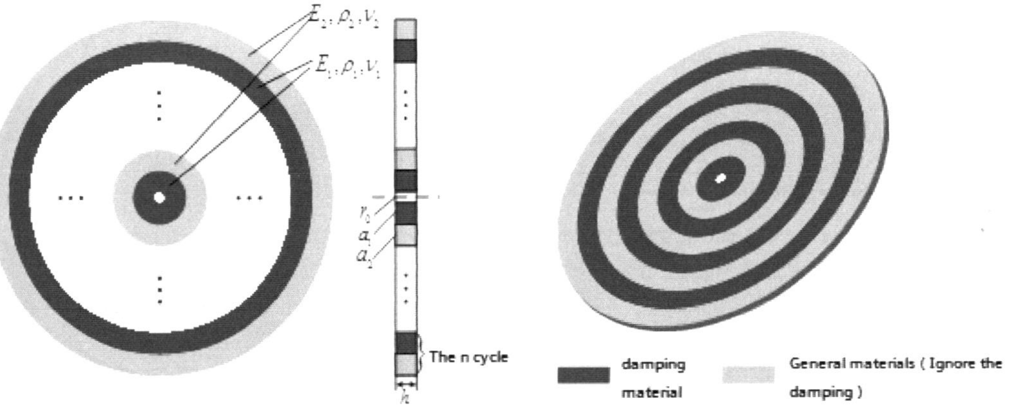

(a) xoy section (b) 3D model

Figure 1. One dimensional radial phononic crystal circular plate.

Damping, usually represented by loss factor β, is a parameter to measure the energy dissipation in the process of elastomer vibration. Because it has a nonlinear functional relationship with strain, stress, temperature, and other variables, it cannot be described by an accurate mathematical model. Therefore, a variety of damping theoretical models has been proposed. In this paper, the viscous damping theory is adopted, and its core expression is:

$$\sigma = E(\varepsilon + \beta \frac{\partial \varepsilon}{\partial t}). \tag{1}$$

Where, $\sigma_d = \beta \frac{\partial \varepsilon}{\partial t}$ represents the damping force, which indicates that the damping force in the solid structure is proportional to the strain velocity. Thus, in the physical equation of bending vibration of the circular plate, equation (1) is introduced to obtain the corresponding stress component expression as follows:

$$\sigma_r = -\frac{E}{1-v^2} z[(\kappa_r + v\kappa_\theta) + \beta(\frac{\kappa_r \partial w}{\partial t} + v\frac{\kappa_\theta \partial w}{\partial t})]. \tag{2}$$

$$\sigma_\theta = -\frac{E}{1-v^2} z[(\kappa_\theta + v\kappa_r) + \beta(\frac{\kappa_\theta \partial w}{\partial t} + v\frac{\kappa_r \partial w}{\partial t})]. \tag{3}$$

$$\tau_{r\theta} = -2z\frac{E}{2(1+v)}(\kappa_{r\theta} + \beta\frac{\kappa_{r\theta}\partial w}{\partial t}). \tag{4}$$

Combined with previous research work [22] and substituting equations (2)~(4), the differential vibration equation of the damped radial circular plate can be derived as follows:

$$\nabla^2\nabla^2 w+\frac{\rho h}{D}\frac{\partial^2 w}{\partial t^2}+\frac{\beta}{D}\frac{\partial w}{\partial t}=\frac{q(r,\theta,t)}{D}. \tag{5}$$

Here and now, in its shape solution $k^4=\omega^2\frac{\rho h}{D}+\omega\frac{\beta}{D}$, D represents bending stiffness. Moreover, it is not difficult to derive the transfer matrix, as shown in formula (6). The specific derivation process is detailed in the literature [22], which will not be repeated here.

$$T_1=R_2^{-1}L_2R_1^{-1}L_1. \tag{6}$$

Where, R_1, L_1, R_2, and L_2 are all square matrices of the second order.

The viscous damping model is used to derive the transfer matrix of a damped radial circular plate. For the analysis in Section 4.2 below, Maxwell mechanical model of viscoelasticity is introduced. First, it is necessary to understand the damping complex modulus model, whose functional relationship is described as:

$$E(\omega)=E_R(\omega)+jE_I(\omega)=E_R(\omega)(1+j\beta). \tag{7}$$

Where E_R and E_I are the energy storage modulus and loss modulus respectively, reflecting the actual energy storage and energy consumption characteristics of viscoelastic materials. It is not difficult to see that they are both functions of frequency. The relation between E_R, E_I, β and frequency ω satisfies the following function:

$$E_R=\frac{E\omega^2\tau^2}{1+\omega^2\tau^2}. \tag{8}$$

$$E_I=\frac{E\omega\tau}{1+\omega^2\tau^2}. \tag{9}$$

$$\beta=\frac{1}{\omega\tau}. \tag{10}$$

Where $\tau=\frac{\eta}{E}$, represents the relaxation time, and η is a coefficient of viscosity.

3. Numerical Analysis of Damped Radial Circular Plates

According to the theoretical analysis results of damping, viscoelastic materials and damping alloys are introduced to construct two types of damped radial phononic crystal circular plates for solving the numerical solution. Material types and material parameters are shown in the following two tables.

Table 1. Material parameters of the first type of damping circular plate.

Material	$\rho\left[kg\cdot m^{-3}\right]$	$E\left[Pa\right]$	ν
Epoxy resin	1800	0.41e10	0.38
STEEL	7850	21e10	0.3

Table 2. Material parameters of the second type of damping circular plate.

Material	$\rho\left[kg\cdot m^{-3}\right]$	$E\left[Pa\right]$	ν
Mn-Cu	8800	10.8e10	0.35
PMMA	1062	0.32e10	0.3333

The period number was selected as $n=4$ and the frequency range was 0~5000Hz. The localization factor of bending vibration of the radial circular plate with different loss factors β was calculated. The relevant geometric parameters are $r_0 = 10mm$, $a_1 = a_2 = 40mm$, $h = 4mm$. The calculation results are shown in figure 2.

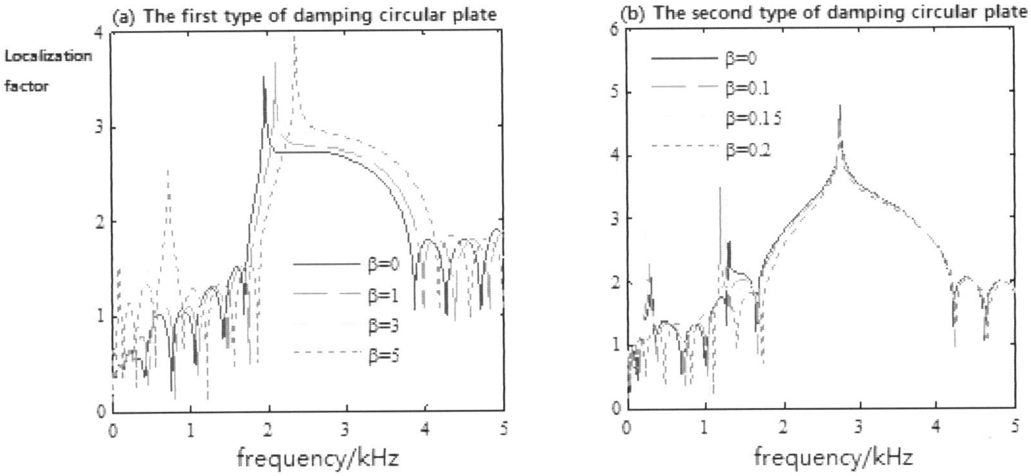

Figure 2. Localization factor of bending vibration of damped thin plate.

It is not difficult to see from figure 2 (a) that, with the increase of β , the band gap initial frequency of the first type of damped thin plate is stable at about 1700Hz, while the cut off frequency is obviously shifted to the high frequency with an amplitude of about 400Hz, which widens the band gap. This shows that the damping effect of viscoelastic materials does have a certain impact on the bandgap characteristics of radial phonon crystal circular plates. Obviously, we can effectively use it in the field of vibration and noise reduction and take targeted measures for noise in certain frequency bands to effectively suppress the propagation of these waves. However, figure 2 (b) shows that the band gap characteristic of the radial circular plate composed of damping alloy is very little affected, which is largely related to the inherent damping characteristic of damping alloy itself, and its small loss factor (ranging from 0.05 to 0.2) makes its role in special occasions and applications far less than that of viscoelastic materials.

4. Simulation Verification of Damped Radial Circular Plate

4.1. Finite Element Analysis of Damped Circular Plate

In order to verify the authenticity and accuracy of the numerical results, it is necessary to carry out finite element simulation analysis. This section adopts the multi-physics field software COMSOL5.3 for analysis, which is different from the relevant settings in literature [22]. In order to objectively reflect the damping effect of materials, we load the "viscoelastic" module in the "linear elastic model" of the solid mechanics' module and set its relaxation time to reflect its damping characteristics.

Two types of damped circular plates are also simulated. In view of the damping problem, according to the relation between viscoelastic damping and relaxation parameters in literature [13], the loss factor β is converted into the corresponding relaxation time τ . The calculation results are shown in figure 3.

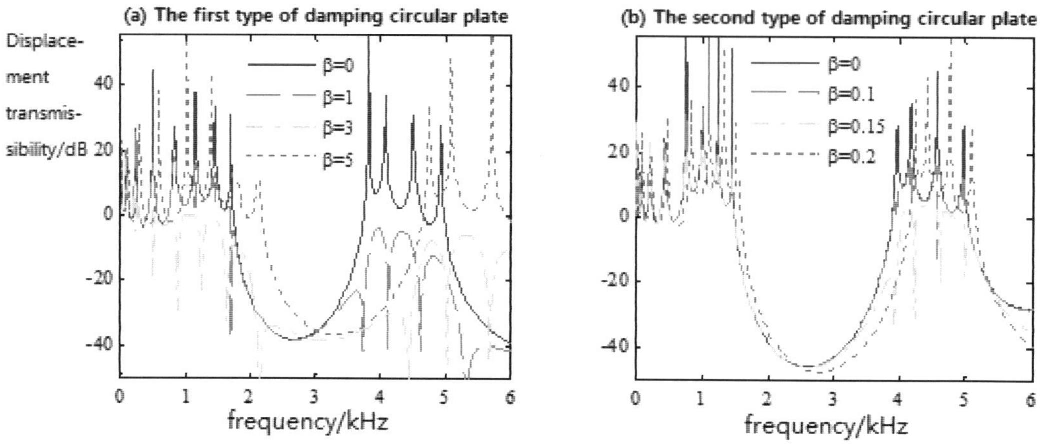

Figure 3. Simulation calculation of damping radial circular plates.

The simulation results in figure 3 have a discrepancy with the numerical calculation results. One of the reasons for this error is that the exact value cannot be obtained when the relaxation time τ corresponding to the loss factor β is calculated. Another reason is that there is a certain gap between the internal calculation model carried by COMSOL and our actual requirements, which leads to some defects in the simulation settings. However, the overall comparison shows that the results of simulation and numerical calculation can confirm each other.

4.2. Frequency Nonlinearity of Elastic Modulus and its Influence

The above starts with the inherent damping characteristics of the material, the viscous damping model is used to investigate the influence of damping on the bandgap characteristics of the radial circular plate, and it is found that the damping effect of viscoelastic material on the band gap is more obvious and more practical than that of the damping alloy. Therefore, in this section, finite element analysis is conducted again from the perspective of frequency nonlinearity of damping and elastic modulus for viscoelastic damping materials (taking epoxy resin as an example).

Taking the relaxation time $\tau = 10^{-2}\,s$, we defined two analytic functions respectively in COMSOL5.3 to represent the energy storage modulus E_R and loss modulus E_I through the formula (8)~(9) and obtained the expression of the complex modulus E^*. The following results are obtained through calculation:

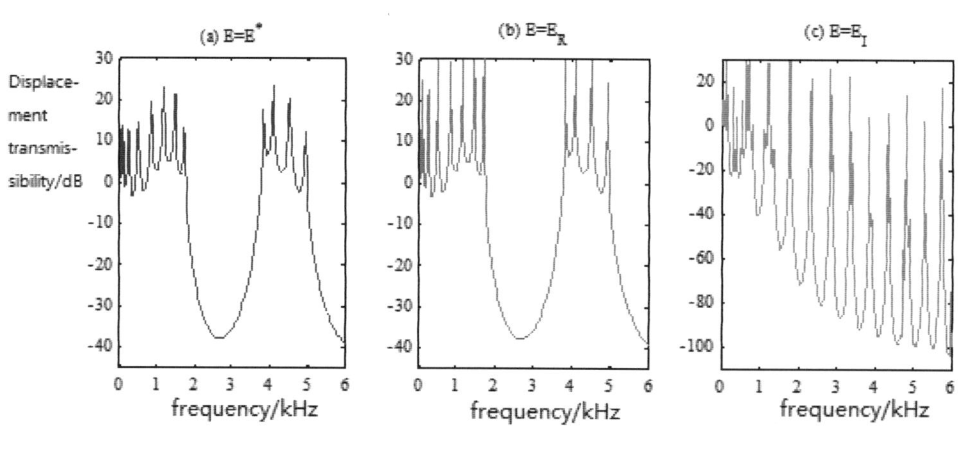

Figure 4. Influence of complex modulus on bandgap characteristics.

It can be seen from figure 4 that the band gap characteristics obtained when the elastic modulus E is complex and constant are quite different. In particular, when E is taken as complex modulus E^* and is taken as storage modulus E_R, the band gap characteristics obtained are very close, which can be clearly seen from figures (a) and (b). When the loss modulus E_I is taken, a series of adjacent narrow band gaps appear in the frequency range 0~6000hz, as shown in (c). This shows that the band gap characteristics of the generalized phononic crystal thin plate are mainly affected by the loss modulus. In order to deeply analyze this influence, the displacement transmissibility when the relaxation time of viscoelastic materials is 10^{-2}, 10^{-3} and 10^{-4} is calculated, and the results are shown in figure 5.

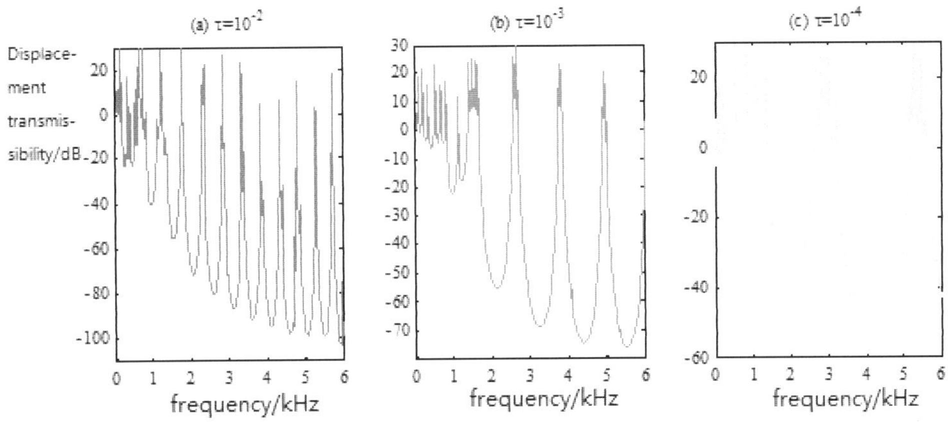

Figure 5. Influence of loss modulus on bandgap characteristics.

It can be clearly seen from figure 5 that the influence law of loss modulus E_I on bandgap under different relaxation time. It is found that the band gap order in the frequency from 0 Hz to 6000Hz decreases with the relaxation time τ decreasing from 10^{-2} to 10^{-4}, that is, the band gap of each order not only has an increasing frequency, but also has an increased bandwidth. Because of the dissipation factor $\beta = 1/\omega\tau$, the smaller the relaxation time, the larger the dissipation factor and the stronger the damping effect. According to figure 5, with the decrease of relaxation time, the loss factor of viscoelastic materials becomes larger and larger, which affects the band gap to move to the high frequency and increases the bandwidth, which is completely consistent with the numerical results and further verifies the accuracy of the theory and numerical simulation.

5. Experimental Verification

The influence of natural damping effect of viscoelastic materials on the bandgap of curved wave in a circular plate of generalized phonon crystal is analyzed by numerical simulation and finite element simulation. In this section, a reasonable and feasible experimental scheme will be designed to fabricate typical damped generalized phonon crystal samples for transverse vibration transfer rate experimental verification.

First of all, it is difficult to directly verify the influence of different damping effects of viscoelastic materials on elastic bandgap characteristics through experiments. It is well known that epoxy resin is a typical viscoelastic material, and it is also a material specially used as the damping layer, while plexiglass is a material whose damping effect can be almost ignored. Meanwhile, it can be seen from table 1 and table 2 that the medium parameters of the two materials are extremely close. Based on the detailed analysis of the influence of various parameters in the reference [22] on the bandgap characteristics, in order to verify the rationality of the experiment, the finite element simulation transmissivity curves in two cases are given first, as shown in figure 6. It can be seen that when the scatterer is epoxy resin with large damping effect, the cut-off frequency of the band gap obviously increases, while the initial position fluctuates a little bit but not much, and the maximum attenuation within the band gap is at the same horizontal line, which is completely consistent with the influence law of damping on the characteristics of the band gap mentioned above.

Therefore, it is reasonable to use these two materials as scatterers and steel as the matrix to construct and process radial phonon crystal samples for experimental analysis.

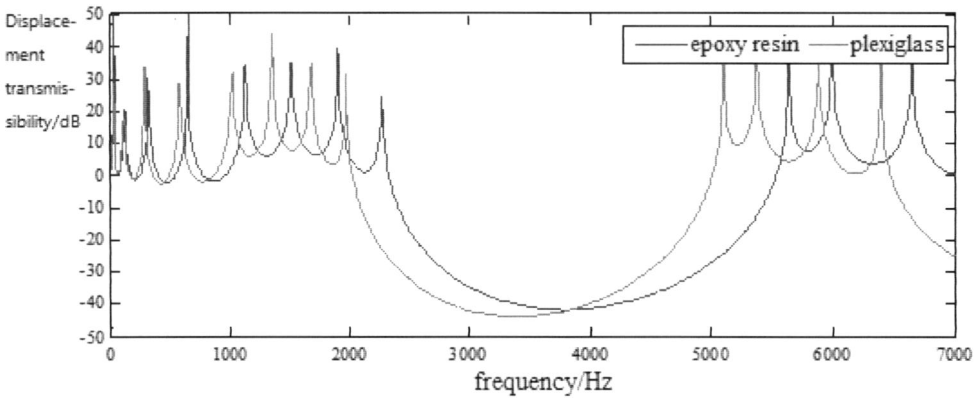

Figure 6. Comparison of the transmissibility curves of the two materials.

Figure 7 shows the layout of the experiment table. The DH5930 dynamic signal test and analysis system designed by Donghua company was used as the signal collection and analysis instrument in the experiment, and the data obtained in the test were saved for subsequent analysis. In the experiment, the damping type circular plate sample is first suspended on the support, and then the micro acceleration sensor is placed on the test point, and the input and output signals are collected by striking the force hammer from the center of the circular plate.

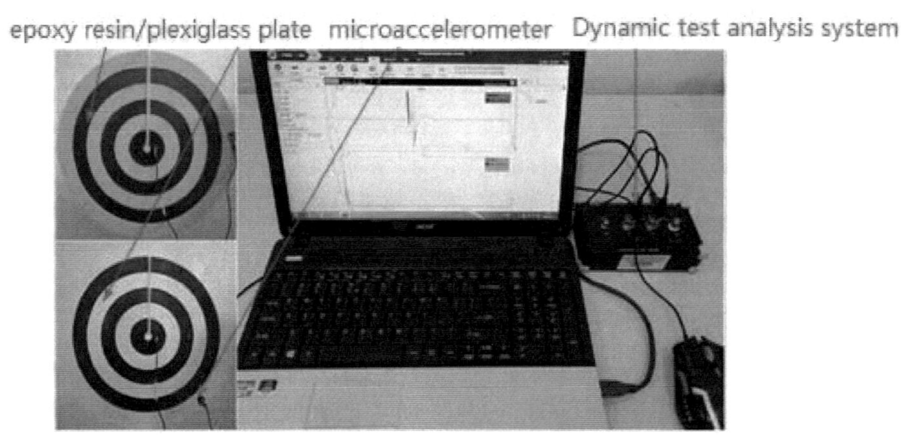

Figure 7. Experimental sample and experimental table layout.

Acceleration vibration data were extracted and the Fourier transform was performed to transform time-domain data into frequency-domain data, and transmissibility curve was obtained after data processing:

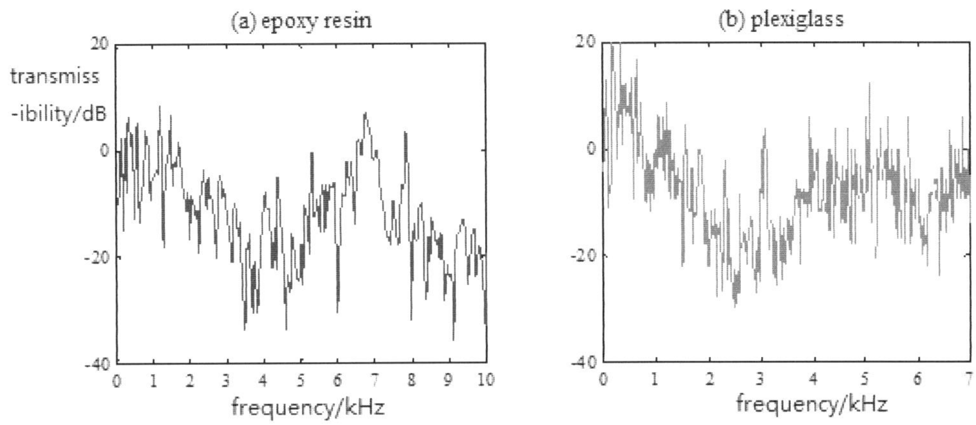

Figure 8. Transmission curve (experimental results).

In figure 8, (a) represents the transmissibility curve measured with epoxy resin, and (b) represents the transmissibility curve measured with plexiglass. By contrast, it is obvious that the starting frequency of the bandgap is basically the same, while the cut-off frequency has a significant change, and the maximum attenuation within the bandgap is the same. This shows that the change of damping effect on the characteristics of the band gap mainly focuses on the cut-off frequency, and the increase of damping has the effect of broadening the bandwidth, and makes the bandgap move to high frequency, which verifies the correctness of the theoretical analysis and simulation results.

6. Conclusion

Through theoretical analysis, numerical analysis and simulation verification of a one-dimensional damped radial phonon crystal circular plate, the following conclusions are drawn:

(1) The damping effect of viscoelastic materials has a great influence on the characteristics of the band gap. With the increase of damping intensity (i.e., the increase of loss factor β), the cut-off frequency of the band gap keeps moving towards the high frequency, while the initial frequency remains basically unchanged, thus broadening the width of the band gap. Obviously, this characteristic of viscoelastic materials can be effectively used in the field of vibration and noise reduction, and targeted measures can be taken for noise in certain frequency bands to effectively inhibit the propagation of these waves.

(2) The damping effect of damping alloy has no significant influence on the bandgap, which is related to the inherent damping characteristic of damping alloy itself. Its small loss factor (ranging from 0.05 to 0.2) makes its role in special occasions and applications far less than that of viscoelastic materials.

References

[1] Du J, Qi P S and Jiang J L 2016 Investigation of Sound Isolation in Composite Structure of Local Resonant Phononic Crystals, Aeronautics and Astronautics Engineering College.

[2] Lee C Y, Leamy M J and Nadler J H 2010 *J. Sound Vib.* Frequency band structure and absorption predictions for multi-periodic acoustic composites, **329(10)** 1809-22.

[3] Munday J N, Bennett C B and Robertson W M 2002 *J. Acoust. Soc. Am.* Band gaps and defect modes in periodically structured waveguides, **112(4)** 1353-58.

[4] Li S B, Chen T N and Xi Y H 2016 Forming Mechanisms of Low-Frequency Complete Band Gaps in Phononic Crystal Plate, School of Mechanical Engineering, **50(12)** 51-7.

[5] Jiang C J and Xiang Y 2018 Analysis and Application of Band Gap Characteristics of Two-dimensional Sonic Crystals, School of Energy and Power Engineering.

[6] Lai Y, et al. 2011 *Nature Mater.* Hybrid elastic solids, **10(8)** 620-24.

[7] Xiao Y, Wen J, Yu D, et al. 2013 *J. Sound Vib.* Flexural wave propagation in beams with periodically attached vibration absorbers: Band-gap behavior and band formation mechanisms, **332** 867-93.

[8] Yu D L, et al. 2006 *Phys. Lett. A* Low frequency torsional vibration gaps in the shaft with locally resonant structures, **348** 410-15.

[9] Wu L Y, Chen L W and Wang C C 2008 *Phys. B Condensed Matt.* Dispersion characteristics of negative refraction sonic crystals, **403(19-20)** 3599-3603.

[10] Achenbach J D and Thau S A 1974 *J. Appl. Mech.* Wave Propagation in Elastic Solids, **41(2)** 544.

[11] Shen X H, et al. 2018 *J. Vib. acoustics* Analysis of Dynamic Behavior of the Finite Elastic Metamaterial-Based Structure With Frequency-Dependent Properties, **140** 3.

[12] Du C Y, Yu D L and Liu J W and Wen J H 2017 *Acta Physica Sinica* Flexural vibration band gaps for a phononic crystal beam with X-shaped local resonance metadamping structure, **66(14)** 140701.

[13] Zhang S W, Wu J H and Liu Z Y 2014 *J. Xi'an Jiaotong Univ.* Damping-Induced Frequency Shifts in Band Structures of One-Dimensional Viscoelastic Phononic Crystal Rods, **3** 22-7.

[14] Xu W k and Wang W 2011 Analysis of Bandgap Characteristic and Effective Parameters of One-dimensional Phononic Crystals Considering Damping.

[15] Hussein M I and Frazier M J 2013 *Acoustic metamater. phononic crystals* Phononic band structures and transmission coefficients: methods and approaches, **173** 329-72.

[16] Wang X G, Shu H S and Zhang L 2019 *J. Appl. Acoustics* Acoustic Insulation Properties of Radial Phononic Crystals, **38** 120-8.

[17] Shu H S, Dong L Q and Li S D 2014 *J. Phys. D: Appl. Phys.* Propagation of torsional waves in a thin circular plate of generalized phononic crystals, **47** 29.

[18] Shi X N, Shu H S and Zhu J 2016 *Int. J. Modern Phys. B* Research on wave band gaps in a circular plate of radial phononic crystal, **30** 1650162.

[19] Shu H S, Wang X G and Liu R 2015 *Int. J. Modern Phys. B* Bandgap analysis of cylindrical shells of generalized phononic crystals by transfer matrix method, **29** 1550176.

[20] Shu H S, Zhang F and Yu H H 2011 *J. Harbin Eng. Univ.* Research on the band gap in a kind of phononic crystal beam with local resonant structure, **9** 1249-53.

[21] Shu H S, Zhang F and Liu S G 2014 *J. Vib. Shock* Vibration band gap of a special rod of phononic crystals, **33** 147-51.

[22] Dong L Q 2015 Vibration characteristics of a one-dimensional generalized phonon crystal circular plate, Master Thesis, Harbin Engineering University.

Acknowledgments

This work was funded by the project (grant number 51875112) supported by the National Natural Science Foundation of China, the project (grant number LBH-Q15029) supported by the Postdoctoral Scientific Research Developmental Fund of Heilongjiang Province of China.

IOP Conf. Series: Materials Science and Engineering **649** (2019) 012008 doi:10.1088/1757-899X/649/1/012008

Zirconium-Based Cladding Coating Technique for Oxidation, Corrosion and Embrittlement Reduction at High-Temperature: An Overview

Mamoun I. A. Sagiroun[1, 2, *] and Cao Xinrong[1]

[1]College of Nuclear Science and Technology, Harbin Engineering University, Harbin, China, 150001
[2]Sudan Atomic Energy Commission, Khartoum, Sudan

sagiroun005@hotmail.com, caoxinrong@hrbeu.edu.cn

Abstract. Fukushima Daichii accident attracted attention to one important factor for nuclear reactors safety – overcoming the high hydrogen emissions under accident conditions. The coating on zirconium-based cladding is one of significant approach to protecting the cladding materials from corrosion, accelerated oxidation at high temperature, embrittlement and (oxygen and hydrogen) uptake at aggressive water-chemical environments inside the core of Light Water Reactors (LWR). This paper presents a compact review of the most common technologies used to apply coating on zr-based cladding superstrates. The paper also discusses the coating materials, substrate preparations and recent advances in deposition techniques.

1. Introduction

Zirconium alloys are extensively used in Light water reactors (LWRs) structural components as fuel claddings due to good physical and mechanical properties which show good performance during normal operation [1, 2]. Many new zirconium alloys cladding materials have been developed by optimizing (Zr-Nb) and (Zr-Sn) alloys series to obtained new zirconium alloys, such as ZIRLO, X5A, and E635 which have corrosion resistance better than Zircaloy-4 [3]. However, the interaction of Zirconium cladding with steam becomes more significant at a temperature greater than 1200 °C which lead to the rapid oxidation of cladding materials [4]. Increasing the production of hydrogen due to zirconium oxidation by high-temperature steam is one of the areas currently being study to make fuel systems inherently safer in the accidents situations, especially after what happened in Fukishma Daichii Reactor [5]. The coating on existing Zr-based claddings is one approach proposed to overcome the rapid oxidation of zirconium alloys at high-temperature and enhance of its corrosion/oxidation resistance during normal operations and accidents condition [6, 7]. In the selection of the coating materials, one must consider a number of important physical and chemical properties such as corrosion resistance, thermal conductivity, thermal stability, adhesion property, high melting point, hydrogen absorption, strength and creep resistance [8]. There are many coating techniques that have been widely investigated for cladding coatings such as thermal spray, physical vapor deposition PVD and chemical vapor deposition CVD [9].

In this review, a brief outline of the previous and current research on the coating materials and their methodology applied to zr-based cladding to protect it against the accelerated oxidation at high-temperature steam is presented.

2. Coating Techniques

Coating technologies are widely used in the industry to enhance the mechanical properties of materials such as resistance to corrosion and wear without a change in the base material. Consequently, this technology was considered by researchers for zirconium alloys cladding materials to slow down oxidation at elevated temperature [10]. One of the most widely used coating/deposition methods is the Chemical Vapor Deposition CVD [11]. Chemical deposition method was used to make a nickel coating on E125 zirconium alloy (Zr-2.5% Nb) which showed the highest efficiency at temperatures up to 800°C [12]. Other commonly used methods for applying coating on the substrate surface layer include Physical Thermal Deposition (PVD) at several variations, evaporative deposition, cathodic arc deposition for instance, pulsed laser deposition, electron beam PVD and sputter deposition. The schematic diagrams in figure 1 show some different coating methods [13].

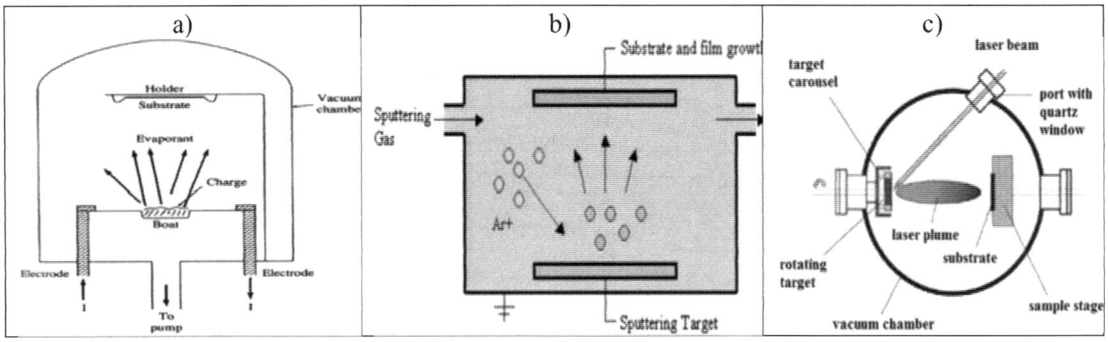

Figure 1. a) Vacuum evaporation system. b) Sputtering deposition system. c) Pulsed laser deposition PLD system[13].

The schematic diagram shown in figure 2 describes the Plasma Immersion Ion Implantation (PIII) method used in implanting Titanium into Zr–2.5Nb alloy which improved the mechanical properties of zirconium substrate [14].

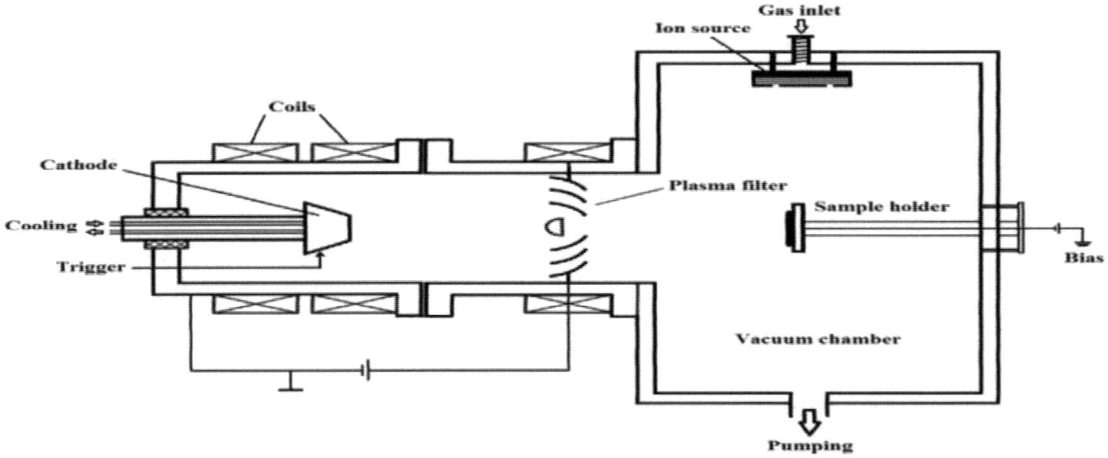

Figure 2. Schematic diagram of the PIII technique [14].

As shown in figure 3, schematic drawings of two different methods of plasma spray (PS) and Laser beam scanning (LBS) used to apply a Si-coated layer on the Zircaloy-4 sample [10].

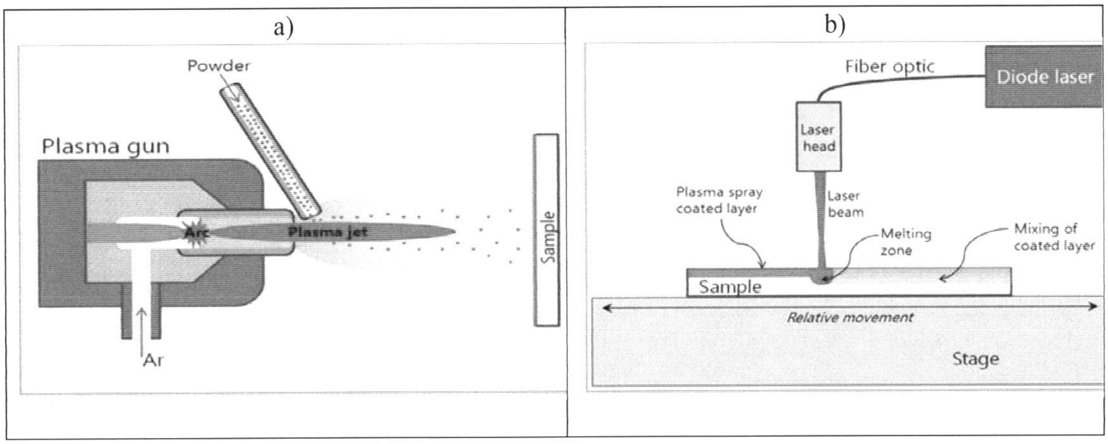

Figure 3. a) plasma spray method (PS) method. b) Laser beam scanning (LBS) method [10].

Thermal spray coating method which melted or heated particles (1–50 μm) and accelerated to high velocities [15]. The High-Velocity Oxygen Fuel (HVOF) spraying method used to protection of (Zr-2.5Nb) zirconium alloy by Cr_3C_2–NiCr coating [16]. Cold spray (CS) process also used for coatings on optimized ZIRLO™ [17].

2.1. Coating Materials on Zr-based cladding

The selection criteria of coating materials on Zr-based cladding as same as that taken into account when selection the elements for alloying of zirconium [8]. Table 1, shows some coating elements properties.

Table 1. Properties for primary elements used as coating materials.

Element	Thermal neutron absorption cross-section (Barns)	Melting temperature (°C)	Thermal Conductivity (W/(m·K))	Oxide	Oxide Melting temperature (°C)
Zr	0.18	1855	22.6	ZrO_2	2715
Ti	6.4	1668	22	TiO_2	1843
Cr	2.9	1850	94	Cr_2O_3	2435
Fe	2.4	1539	80	Fe_2O_3	1565
Ta	21	3017	57	Ta_2O_5	1872
Nb	1.2	2477	53.7	NbO_2	1915
V	4.8	1910	30.7	VO_2	1967
Al	0.22	660	237	Al_2O_3	2044

Many pure elements coating were used to protect the surface of Zr-based cladding, Ti (Titanium) element have good physical properties such as high melting point, thermal conductivity and neutron absorption cross-section which lead to its use as a coating material on z-alloys [18]. Metallic chromium (Cr) has good resistance to high-temperature oxidation which results in it being used as a coating material on Zr-alloys by different coating technologies [19]. Cold spray method is used to apply Cr-coated on Zircaloy-4 material [20]. Also Cr-coating have been deposited on Zircaloy-4 by Arc Ion Plating technology to protect it from high-temperature steam oxidation [21]. AREVA enhanced ATF developments by Cr-coated on M5 cladding [22]. FeCrAl alloys, have good corrosion resistance and high-temperature steam oxidation resistance [23, 24], which protect Zr metal slugs up to 1300 °C when used as coating with Type 310 austenitic stainless steel [25]. FeCrAl coatings deposited on Zircaloy-2 by magnetron sputtering, the samples tested in 700 °C steam, no coating degradation was observed [26]. Nitride ceramic coatings (TiN) also shows good result in corrosion/oxidation resistance when used as a coating on Zr-based claddings [27]. TiN films deposition with Ti-implanted Zr-1Nb alloy by filtered cathodic vacuum arc deposition (FVAD) magnetron sputtering, has been shown to protect (Zr-1Nb) alloy from hydrogen embrittlement [28]. Multilayer (TiN, TiAlN) coatings were deposited onto ZIRLO® cladding by cathodic arc PVD, the corrosion tested in static pure water at (360 oC, 18.7 MPa) for up to 90 days, the results observed best corrosion performance was provided and weight gains six times lower compared by uncoated ZIRLO® [29]. Multilayer Cr-Zr/Cr/Cr-N coatings deposited by the vacuum-arc evaporation technique to protect zirconium alloys from high-temperature oxidation in the air [30]. Some types of carbides, (SiC and SiC/SiC) and their composites were investigated to be used as a cladding material in light water reactors[31, 32]. SiC was deposited on Zircaloy-4 substrates Chemical vapor deposition (CVD) high adhesion strength and corrosion resistance were obtained [30]. Polycrystalline diamond coated applied on Zircaloy-2 fuel tubes used microwave plasma enhanced chemical vapor deposition apparatus with linear antenna delivery, and the test showed that the diamond layer protect the zirconium against hot steam oxidation[33]. The surface of zirconium alloy (ZIRLO) was also protected by Nanocrystalline diamond layer against oxygen and hydrogen uptake [34].

3. Conclusion

- Different types of coating materials such as single, multilayer, metallic and ceramic materials have been used, and they showed good results for protecting the zr-based cladding from high oxidation, corrosion and Oxygen/hydrogen uptake.
- The coatings on zr-based cladding were developed to protect the base materials from water side corrosion and rapid oxidation at high temperature, so it is important the coating layers should be highly stable and well adherent/coherent with the base material.
- One of the major challenges to the structural integrity of coating layers is strain mismatch between coated layer and zircaloy during increasing burnup, due to their different irradiation induced axial growth and creep deformation [35].

References

[1] Bragg-Sitton S N 2012 *US Dep. Energy* Advanced LWR Nuclear Fuel Cladding System Development : Technical Program Plan, no. Idaho National Laboratory, 114.
[2] Azevedo C R F 2011 *Eng. Fail. Anal.* Selection of fuel cladding material for nuclear fission reactors, **18(8)** 1943-62.
[3] Sun C, Yang Z and Wu Z 2018 *World J. Nucl. Sci. Technol.* Study on Corrosion Resistance of N36 Zirconium Alloy in LiOH Aqueous Solution, **8(2)** 30-7.
[4] Cheng T, Keiser J R, Brady M P, Terrani K A and Pint B A 2012 *J. Nucl. Mater.* Oxidation of fuel cladding candidate materials in steam environments at high temperature and pressure, **427(1-3)** 396-400.
[5] Bragg-Sitton S N 2014 *Nucl. News* Development of advanced accident-tolerant fules for commercial LWRs, 83-91.
[6] Kim H G, Kim I H, Jung Y I, Park D J, Yang J H and Koo Y H 2016 *Top Fuel 2016* Development of Surface Modified Zr Cladding by Coating Technology for ATF, 1157-63.
[7] Kim H G, et al. 2017 *Water Reactor Fuel Perform. Meeting. Korea* Progress of surface modified Zr cladding development for ATF in Korea.
[8] Baczynski J R 2014 High temperature steam oxidation of titanium-coated Zircaloy-2 and Titanium-Zirconium Alloys, 92.
[9] Duan Z, et al. 2017 *Nucl. Eng. Des.* Current status of materials development of nuclear fuel cladding tubes for light water reactors, **316** 131-50.
[10] Kim H G, Kim I H, Park J Y and Koo Y H 2014 *Zircon. Nucl. Ind. 17th Vol.* Application of Coating Technology on Zirconium-Based Alloy to Decrease High-Temperature Oxidation, 346-69.
[11] Malandrino G 2009 *Angew. Chemie Int. Ed.* Chemical Vapour Deposition. Precursors, Processes and Applications. Edited by Anthony C. Jones and Michael L. Hitchman., **48(41)**

7478-9.
[12] Valeeva A K, Valeev I S, Valiakhmetov O R, Zhemchuzhnikova D A, Markushev M V and Mulyukov R R 2012 *Inorg. Mater. Appl. Res.* Ni-based protective-lubricant coatings for zirconium alloys, **3(3)** 226-30.
[13] Shahidi S, Moazzenchi B and Ghoranneviss M 2015 *Eur. Phys. J. Appl. Phys.* A review-application of physical vapor deposition (PVD) and related methods in the textile industry **71(3)** 31302.
[14] Kashkarov E B, Nikitenkov N N, Syrtanov M S, Sutygina A N, Shulepov I A and Lider A M 2016 *Appl. Surf. Sci.* Influence of plasma immersion titanium implantation on hydrogenation and mechanical properties of Zr-2.5Nb, **370** 142-8.
[15] Herman H, Sampath S and McCune R 2000 *MRS Bulletin* Thermal Spray: Current Status and Future Trends, **25(7)** 17-25.
[16] Jin D, et al. 2016 *Surf. Coatings Technol.* A study of the zirconium alloy protection by Cr 3 C 2 -NiCr coating for nuclear reactor application, **287** 55-60.
[17] Yeom H, Maier B, Johnson G, Dabney T, Walters J and Sridharan K 2018 *J. Nucl. Mater.* Development of cold spray process for oxidation-resistant FeCrAl and Mo diffusion barrier coatings on optimized ZIRLO^{TM}, **507** 306-15.
[18] Kashkarov E B, et al. 2018 *Int. J. Hydrogen Energy* Hydrogen absorption by Ti-implanted Zr-1Nb alloy, **43(4)** 2484-91.
[19] Brachet J C, et al. 2015 *Top Fuel* On-going studies at CEA on chromium coated zirconium based nuclear fuel claddings for enhanced Accident Tolerant LWRs Fuel, 31-8.
[20] Ševeček M, et al. 2018 *Nucl. Eng. Technol.* Development of Cr cold spray–coated fuel cladding with enhanced accident tolerance, **50(2)** 229-36.
[21] Park J H, Kim H G, Park J Y, Jung Y I, Park D J and Koo Y H 2015 *Surf. Coatings Technol.* High temperature steam-oxidation behavior of arc ion plated Cr coatings for accident tolerant fuel claddings, **280** 256-9.
[22] Bischoff J, et al. 2018 *Nucl. Eng. Technol.* AREVA NP's enhanced accident-tolerant fuel developments: Focus on Cr-coated M5 cladding, **50(2)** 223-8.
[23] Park D J, Kim H G, Park J Y, Jung Y I, Park J H and Koo Y H 2015 *Corros. Sci.* A study of the oxidation of FeCrAl alloy in pressurized water and high-temperature steam environment, **94** 459-65.
[24] Pan D, Zhang R, Wang H, Lu C and Liu Y 2016 *J. Alloys Compd.* Formation and stability of oxide layer in FeCrAl fuel cladding material under high-temperature steam, **684** 549-55.
[25] Terrani K A, Parish C M, Shin D and Pint B A 2013 *J. Nucl. Mater.* Protection of zirconium by alumina- and chromia-forming iron alloys under high-temperature steam exposure, **438(1-3)** 64-71.
[26] Zhong W, Mouche P A, Han X, Heuser B J, Mandapaka K K and Was G S 2016 *J. Nucl. Mater.* Performance of iron-chromium-aluminum alloy surface coatings on Zircaloy 2 under high-temperature steam and normal BWR operating conditions, **470** 327-38.
[27] Zhang X, et al. 2014 *J. Nucl. Mater.* Superior corrosion resistance properties of TiN-based coatings on Zircaloy tubes in supercritical water, **451(1-3)** 346-51.
[28] Kashkarov E B, et al. 2018 *Appl. Surf. Sci.* Hydrogenation behavior of Ti-implanted Zr-1Nb alloy with TiN films deposited using filtered vacuum arc and magnetron sputtering, **432** 207-13.
[29] Alat E, Motta A T, Comstock R J, Partezana J M and Wolfe D E 2016 *J. Nucl. Mater.* Multilayer (TiN , TiAlN) ceramic coatings for nuclear fuel cladding, **478** 236-44.
[30] Kuprin A S, et al. 2015 *J. Nucl. Mater.* Vacuum-arc chromium-based coatings for protection of zirconium alloys from the high-temperature oxidation in air, **465** 400-6.
[31] Avincola V A, Grosse M, Stegmaier U, Steinbrueck M and Seifert H J 2015 *Nucl. Eng. Des.* Oxidation at high temperatures in steam atmosphere and quench of silicon carbide composites for nuclear application, **295** 468-78.
[32] Singh G, Sweet R, Brown N R, Wirth B D, Katoh Y and Terrani K, 2018 *J. Nucl. Mater.* Parametric Evaluation of SiC/SiC Composite Cladding with UO2 Fuel for LWR Applications: Fuel Rod Interactions and Impact of Nonuniform Power Profile in Fuel Rod, **499** 155-67.

[33] Kratochvílová I, et al. 2014 *J. Mater. Process. Technol.* Nanosized polycrystalline diamond cladding for surface protection of zirconium nuclear fuel tubes, **214(11)** 2600-5.
[34] Škarohlíd J, et al. 2017 *Sci. Rep.* Nanocrystalline diamond protects Zr cladding surface against oxygen and hydrogen uptake: Nuclear fuel durability enhancement, **7(1)** 1-14.
[35] Lee Y, Lee J I and NO H C 2017 *Nucl. Eng. Technol.* Mechanical analysis of surface-coated zircaloy cladding, **49(5)** 1031-43.

Acknowledgments
Thank the College of Nuclear Science and Technology, Harbin Engineering University, China and China Scholarship Council (CSC) for supporting this research, and we would also like to thank Mr. Ayodeji Abiodun and Mr. Anuor Ayoub their help.

Vibration Characteristics of Elastic Metamaterial Rod

Ziyue Zheng[1], Haisheng Shu[1], Shuowei A[1], Di Mu[1], Lei Zhao[1] and Yuqiang Su[1]

[1]Harbin Engineering University, College of Mechanical and Electrical Engineering, 150001, Harbin, China

shujs@hrbeu.edu.cn

Abstract. This paper investigates the free vibration and steady forced vibration of elastic metamaterial rods. Firstly, the elastic metamaterial rod is equivalent to the effective medium rod based on the effective medium theory. Then the free vibration and steady forced vibration method of the effective medium rod are established from the classical theory, and with this method we can explored the difference characteristics between the elastic metamaterial rod and the natural media rods. The results show that the effective medium rod has some strange phenomenon such as the same shape of different natural frequencies, the accumulation of natural frequencies near the lower band and the loss of natural frequency. This work is expected to help the vibration control of the elastic metamaterial for finite structure.

1. Introduction

In 2000, Liu took the lead in proposing a local resonance phononic crystal structure, and pointed out that since the scatterer has a local resonance of the mode "spring-mass", the structure can be regarded as a homogenous structure, which has frequency and negative density under the long wave limit [1]. Subsequently, a large number of scholars have designed different unit cell structures and initiated different types of resonance modes (monopoles, dipoles, quadrupoles, etc.) of the scatterers to obtain shear modulus, bulk modulus, etc., which have elastic/acoustic metamaterial structure [2-7]. Thanks to these strange characteristic different from those of natural materials, scholars applied it to low frequency vibration [8-10], negative refraction [2, 5, 11], acoustic/elastic wave hiding [12, 13], seismic wave protection [14-16, 27] and other aspects.

In recent years, due to the development of effective medium theory [2, 17-22, 29], more and more scholars have begun to use this method to explore the propagation characteristics of acoustic/elastic waves in infinite or semi-infinite elastic metamaterial [3, 16, 23-26]. The effective medium theory is considered that under the long-wave limit, when the wavelength in the matrix of the elastic metamaterial is much larger than the lattice constant, the structure can be equivalent to the homogeneous structure macroscopic which material parameter is frequency changed (equivalent density ρ_{eff}, equivalent shear modulus μ_{eff}, the equivalent bulk modulus K_{eff}). Compared with the modeling of the unit cell level, this not only greatly simplifies the calculation, but also facilitates people to understand the propagation behavior of long waves in elastic metamaterials. The most classic is the interpretation of the band gap and negative refraction characteristics [5, 19]. Williams [25] and others analytically calculated the equivalent density of the multi-vibrator metamaterial beam to explain the propagation behavior of the Lamb wave. R. Zhu [26] verified the anisotropy of equivalent density by experimental method, and discussed the propagation behavior of waves in anisotropic elastic metamaterial plates. In addition, based on the effective medium theory, our group [16, 24] analyzed the Lamb wave in the elastic metamaterial layer and the propagation behavior of the SH wave in the associated half space, and predicted the possible traveling wave components. Using the effective medium theory, the above work is a good analysis of the propagation of elastic waves in an infinite or semi-infinite elastic metamaterial medium, which can be applied to occasion where the wavelength is much smaller than the propagation size, such as seismic wave protection [14].

However, it should be noted that when the concept of elastic metamaterials is introduced into practical applications, the corresponding structures are generally waveguide structures of finite size, and the main concern is the vibration characteristics of these special structures.

In this paper, we use the effective medium theory to equate an elastic metamaterial rod into an effective medium rod. Starting from the classical theory (one-dimensional wave equation), the free vibration and forced vibration analysis methods of the effective medium rod are established, and the free vibration and forced vibration characteristics of the elastic metamaterial rod are explored by this method. When investigating free vibration, due to the frequency change and the negative value of the equivalent density, the finite elastic metamaterial rod has significantly different modal characteristics from the natural medium rod, including the same mode shape at different natural frequency points, the dense distribution of natural frequencies near the band edge and the absence of natural frequencies and modes. The analysis of the forced vibration characteristics further demonstrates the effectiveness of free vibration analysis.

2. Analysis of Free Vibration of EM Rod

Under the condition of two-dimensional plane stress, this paper adopts the elastic metamaterial rod structure shown in figure 1.a, which is made up of a single cell structure as shown in figure 1.b through a 10*3 array. The material and structural parameters are seen in table 1. With the help of the "Feel and Response" method [2], the unit cell equivalent parameters are calculated as shown in figure 1.b. The equivalent Young's modulus E_{eff} can be obtained by $E(\omega) = \dfrac{9KG}{3K+G}$.

It can be seen from figure 1.b that in 0-2500 Hz, the equivalent density varies greatly due to the excitation of the scatterer dipole mode, and a negative value occurs in the 570-1258 Hz (gray region). However, the equivalent shear modulus and the equivalent bulk modulus hardly change, so this paper will focus on the frequency change of the equivalent density. At the same time, in order to facilitate the subsequent analysis, we show the dispersion relationship of the longitudinal wave in such a rod as shown in figure 1.c, which satisfies: $\omega = c_b k = \sqrt{E_{eff}/\rho_{eff}} \cdot k$. Due to the frequency variation of the equivalent density, the longitudinal wave is dispersed, and the energy band curve is decomposed into two branches, the sound branches and the optical branch which are separated by a band gap (grey area), and the band gap interval corresponds to the negative density interval [19, 4]. The band structure can also be given by FEM as shown in figure 1.c. This result shows good agreement with the theoretical calculation results of equivalent medium, which confirms the validity of the effective medium theory.

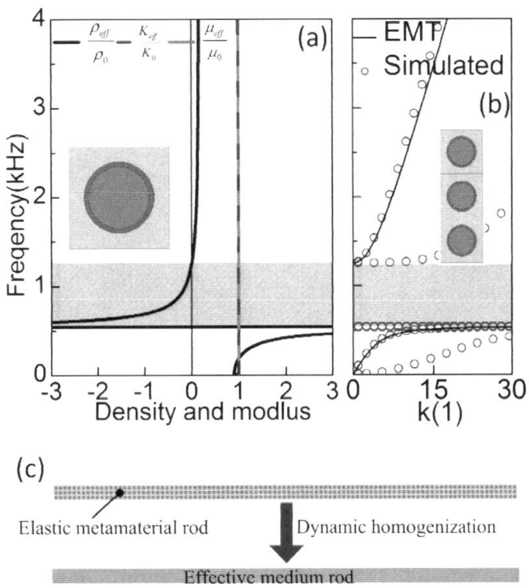

Figure 1. (a) composition of unit cell and equivalent parameters (b) 1 * 3 primitive cell and its band structure (c) equivalent process.

Table 1. Unit cell material parameter.

component	ρ (kg/m^3)	E(Pa)	G(Pa)
matrix	1180	0.435e10	0.159e10
scatterer	11600	0.408e11	0.149e11
layer	1300	1.175e5	4e4

Table 2. Unit cell structure and parameters.

a	R	r
10	3.5	3

Next, the above effective medium rod free vibration is analyzed. It is noted that at this time the rod is no longer a Cauchy medium, and its material parameters, especially the equivalent density, have a strong frequency change, especially a negative value in the band gap, then it is necessary to establish a free vibration analysis method of effective medium rod. Following the classical theory, the effective medium rod governing equation can be written as [31]:

$$\frac{\partial^2 u}{\partial x^2}-\frac{1}{c_b^2(\omega)}\frac{\partial^2 u}{\partial t^2}=0 \qquad 0\le x\le l \qquad -\infty<t<-\infty \tag{1}$$

Where l is the length of the rod, $c_b(\omega)$ is the velocity of longitudinal wave in the rod, and the magnitude is $\sqrt{\frac{E(\omega)}{\rho(\omega)}}$, where the velocity of longitudinal wave, the elastic modulus and the density are both a function of the circular (angle) ω frequency. We find that when $\frac{1}{c_b^2(\omega)}$ is a positive number, this variable ω has no effect on the solution of equation (1). Therefore, according to the classical theory method, let the standing wave solution be brought into equation (1), we can get:

$$\frac{X''}{X} = \frac{1}{c_b^2(\omega)} \frac{T''}{T} = -k(\omega)^2 \tag{2}$$

When considering the boundary condition of the effective medium rod as free at both ends, to put the boundary condition with zero stress at both ends into equation (2), and the non-zero solution is obtained to obtain the natural frequency distribution as follows:

$$\frac{\omega}{n} = \frac{\pi \sqrt{E(\omega) / \rho(\omega)}}{l}, n = 1, 2, 3, ... \tag{3}$$

In particular, when the equivalent density of figure 1 also exhibits a negative value in the band gap frequency, to solve the governing equation (1), it is noted that $\frac{1}{c_b^2(\omega)}$ is also a negative value.

If: $k \neq 0$, for the second-order partial differential equation, it can also be solved by the standing wave method, and its solution is:

$$u = X\cdot T = (Ae^{-k^* x} + Be^{k^* x})\cdot(Ce^{j\omega t} + De^{-j\omega t}) \tag{4}$$

Where k^* is a non-negative real number, satisfied $k^* = i\cdot k$. Bring the free boundary conditions at both ends into equation (4):

$$\begin{bmatrix} -k^* & k^* \\ -k^* e^{-k^* l} & k^* e^{k^* l} \end{bmatrix} \cdot \begin{bmatrix} A \\ B \end{bmatrix} = 0 \tag{5}$$

If we have a non-zero solution in equation (5), k^* is zero only. Obviously, this contradicts the assumptions, so we give up.

If: $k=0$, then $\omega = c_b k = 0$,means that there is no natural frequency in the effective medium rod. At this time, the right side of equation (2) is 0, then the solution of the differential equation is:

$$u = X\cdot T = (A + Bx)\cdot(C + Dt) \tag{6}$$

Bring the boundary conditions of free at both ends, we can get $B = 0$ in equation (6). From $t \to \infty$, the above formula u must be a finite number to have physical meaning, and get: $D = 0$ .Thereby you can get:

$$u = X\cdot T = A\cdot C = u_0 \tag{7}$$

IOP Conf. Series: Materials Science and Engineering **649** (2019) 012009 doi:10.1088/1757-899X/649/1/012009

The free vibration analysis method of the effective medium rod is given above. Due to the existence of the variables in the governing eqaution (1), the equation (3) characterizing the natural frequency distribution and the corresponding modal shape is obviously different from the classical theoretical results. Also, the above also gives the analysis of free vibration when the density is negative, and this case has never been covered by classical theory. Next, we will use the above analysis method to analyze the free vibration characteristics of the effective medium rod shown in figure 1.

As shown in figure 1, ρ_{eff} varies with frequency. When approaching 570 Hz, that is, the lower band edge f_0^- of the band gap, ρeff will rise sharply with frequency; while near 1258 Hz, that is, the upper band edge f_1^+ of the band gap, ρ_{eff} will rise relatively slowly from 0; Between f_0^+ and f_1^- ρ_{eff} is negative.

First consider ρ_{eff} as positive and bring its value into equation (3). For the convenience of expression, let the left side of the equation (3) equal sign y_1^n and the right side y_2, and figure 2 gives the function image of y_1^n and y_2. The abscissa of the intersection of the two functions is the natural frequency, and the corresponding mode is $\cos(\frac{n\pi}{l}x)$, and n represents the nth mode. For comparison, the natural frequency and modal state in the natural medium are also given, where E and ρ are taken when $\omega \to 0$

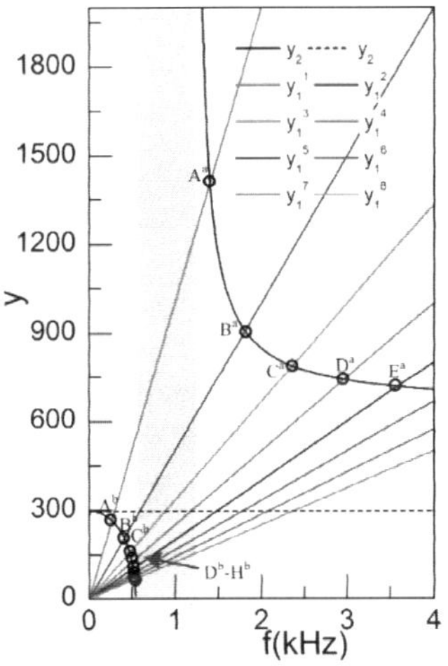

Figure 2. the function image of y_1^n and y_2.

We know that there is only one intersection of y_2 and y_1^n in the natural media rod as shown in figure 2, which means that a natural frequency corresponds to a modality $\cos\frac{n\pi x}{l}$. However, for the effective medium rod, due to the frequency variation of its density, y_2 splits into two upper and lower branches, resulting in an intersection of y_1^n and y_2 before and after band gap, which means that the same mode will correspond to two natural frequencies. And the two natural frequencies are the nth order before the band gap and the nth order after the band gap, which is obviously different from the natural dielectric rod. In addition, in figure 2 we also observe when the frequency approaches the lower band edge f_0^-, $y_2 \to 0$ $y_2^{'} \to \infty$, then the first intersection with this will be clustered nearby f_0^-, and the intersection spacing becomes narrower and narrower. That is to say, the effective medium rods in the vicinity of f_0^- will have an infinite number of natural frequencies and their modal density will increase.

In fact, the essence of the above-mentioned peculiar phenomenon is derived from the modulation of the scatterer dipole resonance on the longitudinal wave in the rod, and the dispersion is given in figure 1. On the one hand, it can be seen that for longitudinal wave and traveling waves of the same frequency, longitudinal wave propagation of two wavelengths is allowed in the elastic metamaterial due to modulation. Since the modality can be seen as a standing wave form and consists of a set of traveling waves propagating in the opposite direction of the same frequency, and the modal shape is determined only by the wavelength. Then, two sets of traveling wave with the same wavelength and different frequencies in the elastic metamaterial rod naturally produce the same modal shape but different modal frequencies. On the other hand, near the vicinity of f_0^- (area A of figure 1), the dispersion curve tends to be a straight band. This means that the two different wavelengths of the longitudinal waves will have a very small difference in frequency, and this difference will become smaller and smaller as the frequency approaches f_0^-. When two sets of traveling waves with different wavelengths and almost the same frequency form a mode, these modal frequencies will naturally be close, and as the frequency approaches f_0^-, the modal frequency interval becomes smaller, that is, the modal density increases.

When the frequency is between f_0^+ and f_1^- , ρ_{eff} is negative, it can be seen from the above analysis that there is no natural frequency in the effective medium rod, and equation (7) means that the motion state of the rod can be regarded as the rigid body is still and always at the initial displacement u_0. In fact, this solution is in line with the physical nature of negative density. From the unit cell level, when the dipole resonance of the scatterer in the rod is excited, the displacement in the rod matrix is offset to zero by the displacement in the scatterer at the band gap frequency (negative density interval), that is, at this time, there is no mode in the matrix that provides vibration.

3. Analysis of Steady Forced Vibration

Considering that the structure will always be subjected to a certain dynamic load rather than free vibration under actual conditions, we will discuss the forced vibration of the effective medium rod in steady state. Similarly, it is first necessary to establish an effective medium, a forced vibration method for a non-cauchy dielectric rod. Compared with the classical theory, c_b^2 of the governing equation (1) has frequency change and negative values, but we find that this does not affect the solution, that is, the steady solution of equation (1) can still be written as:

$$u(x,t) = u^r + u^l = (Ae^{-jk(\omega)x} + Be^{jk(\omega)x})e^{j\omega t} \tag{8}$$

The wave vector is $k = \omega / c_b(\omega)$. A, B are determined by the boundary conditions of the rod. Consider the boundary condition of the rod in the second section, where the left end is excited by the harmonic displacement $u = u_0 \cdot e^{j\omega t}$, and when the right end is free, the equation (8) has: $A = \dfrac{u_0 \cdot e^{jkl}}{e^{jkl} + e^{-jkl}}$, $B = \dfrac{u_0 \cdot e^{-jkl}}{e^{jkl} + e^{-jkl}}$. The displacement transfer rate from the left end to the right end of the rod $Trans$ can be obtained by the following formula.

$$Trans = 20 \cdot \lg \frac{u_l}{u_0} \tag{9}$$

Where, u_l is the displacement amplitude at the right end of the rod can be obtained from equation (8). Figure 3a shows the effective medium rod displacement transfer rate curve in the 1 to 2500 Hz band.

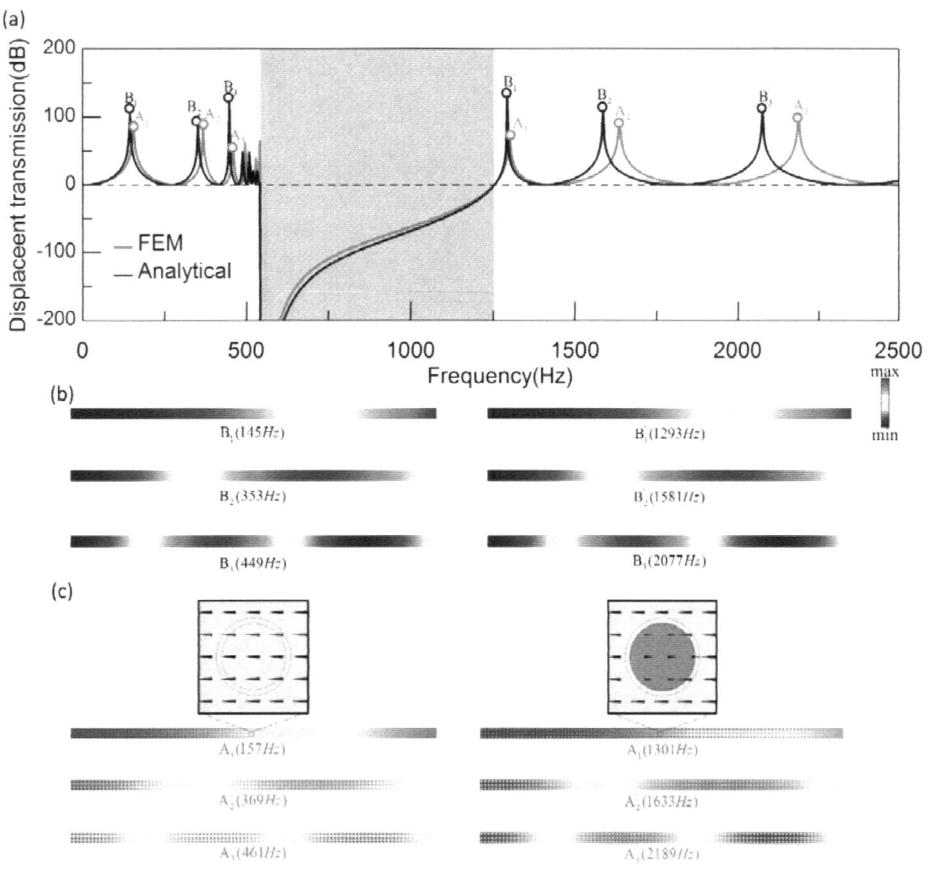

Figure 3. (a) Forced vibration transfer rate curve (b) Theoretical analysis of the peak displacement field distribution (c) Simulation of the peak displacement field distribution and the corresponding unit cell displacement field magnified view, the arrow represents the displacement vector.

It is observed in the transfer rate curve of figure 3a that a large number of formants are enriched near the lower band edge and the interval is smaller and smaller; the transfer rate in the band gap frequency is very small. Using the equation (8) to calculate the displacement field at the formats in and after the band gap, it is found that the existence is similar to that in the free vibration analysis, that is, the displacement field of the nth formant before the band gap and after the band gap remain consistent. In fact, from the point of view of the modal superposition method, these steady forced vibration characteristics of the rod are actually another manifestation of its free vibration characteristics. Since the effective medium rod is concentrated at the natural frequency near the lower band edge, when the excitation is close to these concentrated natural frequencies, the format generated by the rod response is naturally concentrated; since There is not a natural frequency in the negative density interval (band gap frequency) of the effective medium rod, its motion form can be regarded as the static state of the rigid body. When the excitation frequency is in the negative density interval, the motion state of the rod is dominated by the stationary state, which leads to the decrease of the transfer rate; the band gap has the same modal shape successively. The mode is naturally excited to exhibit a similar displacement field response. In addition, it should be pointed out that due to the relationship between the steady forced vibration of the effective medium rod and the free vibration, the forced vibration characteristic results also confirm the reliability of the above free vibration analysis.

In order to further verify the effectiveness of the above-mentioned effective meduim rod forced vibration analysis, we used COMSOL to model the elastic metamaterial rod at the unit cell level and perform steady state calculation. Applying a simple harmonic displacement excitation $u = u_0 e^{j\omega t}$ to the left end of the rod which direction is x as shown in figure 1c, u_0 is the displacement amplitude, ω is the circle (angle) frequency, the other surfaces are free. Take the average displacement of the line at the right end of the pickup rod u_l, the displacement transfer rate curve can be obtained by using equation (9) as shown in figure 3a.

We can see that the transfer rate curve of the effective medium rod and the elastic metamaterial rod modeled from the unit cell level shows a good agreement at the low frequency band, especially at the band gap position. However, we still need to point out that this analysis process is based on the dynamic homogenization theory under the long-wave limit and cannot describe the unit cell motion state. For example, in figure 3c, although the corresponding one of the first formants before and after the band gap is macroscopically consistent, the displacement field on a single unit cell (microscopic) exhibits an inconsistency similar to that in the free vibration analysis. Moreover, the accuracy of this algorithm depends to a large extent on the accuracy of the effective medium theory. It can be seen from figure. 3a that the transfer rate curve deteriorates after 1500 Hz, which is due to the boundary effect and high frequency factors leading to the error of the equivalent medium theory [20]. Even so, we still need to emphasize that the above method can still provide the ideal accuracy for the forced vibration of the elastic metamaterial rod at low frequencies (maximum error of peak position is 4.3%).

4. Conclusion

In this paper, we use the effective medium theory to analyze the free vibration and steady forced vibration analysis method of a finite-size effective medium rod from the governing equation of the rod (one-dimensional wave equation). And using this method, explore the vibration characteristics of the elastic metamaterial rod different from the natural medium rod. The theoretical results at the effective medium level together with the simulation results at the unit cell level show that: 1. Due to the dipole resonance for the modulation of the longitudinal waves in the rod, the nth order modal shape of the elastic metamaterial rod before and after the band gap is macroscopically uniform. In the effective medium theory, different natural frequencies correspond to the same natural mode; 2, the elastic metamaterial rod does not have a natural frequency in the band gap frequency interval (negative density interval). Its motion form is similar to a stationary rigid body; 3. The natural frequency of the metamaterial rod is concentrated at the lower band edge, and the mode density is continuously increased. In this paper, an analytical method for free vibration and steady forced vibration of an effective medium rod is established. This method can be extended to the elastic metamaterial finite structures such as plates, shells and films, and it is expected that this work will help people to understand and solve the dynamic response of elastic metamaterial finite structures from the half-wavelength scale and more fully utilize its characteristics to control the vibration displacement field and vibration.

References

[1] Liu Z, Zhang X, Mao Y, et al. 2000 *Sci.* Locally resonant sonic materials, **289(5485)** 1734-36.
[2] Liu X N, Hu G K, Huang G L, et al. 2011 *Appl. Phys. Lett.* An elastic metamaterial with simultaneously negative mass density and bulk modulus, **98(25)** 251907.
[3] Lai Y, Wu Y, Sheng P, et al. 2011 *Nature Mater.* Hybrid elastic solids, **10(8)** 620.
[4] Yang Z, Mei J, Yang M, et al. 2008 *Phys.* Membrane-type acoustic metamaterial with negative dynamic mass, **101(20)** 204301.
[5] Wu Y, Lai Y and Zhang Z Q 2011 *Physic. Rev. Lett.* Elastic metamaterials with simultaneously negative effective shear modulus and mass density, **107(10)** 105506.
[6] Oh J W, Kwon Y E, Lee H J and Kim Y Y 2016 *Sci. Rep.* Elastic metamaterials for independent realization of negativity in density and stiffness, **6** 23630.
[7] Li J, Chan C T 2004 *Physic. Rev. E* Double-negative acoustic metamaterial, **70** 055602.
[8] Zhu R, Liu X N, Hu G K, et al. 2014 *J. Sound Vib.* A chiral elastic metamaterial beam for broadband vibration suppression, **333(10)** 2759-73.
[9] Liu X N, Hu G K, Sun C T, et al. 2011 *J. Sound Vib.* Wave propagation characterization and design of two-dimensional elastic chiral metacomposite, **330(11)** 2536-53.

[10] Xiao Y, Wen J, Yu D, et al. 2013 *J. Sound Vib.* Flexural wave propagation in beams with periodically attached vibration absorbers: Band-gap behavior and band formation mechanisms, **332(4)** 867-93.
[11] Feng L, Liu X P, Lu M H, et al. 2006 *Physic. Rev. Lett.* Acoustic backward-wave negative refractions in the second band of a sonic crystal, **96(1)** 014301.
[12] Farhat M, Guenneau S and Enoch S 2009 *Physic. Rev. Lett.* Ultrabroadband elastic cloaking in thin plates, **103(2)** 024301.
[13] Stenger N, Wilhelm M and Wegener M 2012 *Physic. Rev. Lett.* Experiments on elastic cloaking in thin plates, **108(1)** 014301.
[14] Andrea C, Philippe R, Sebastien G, et al. 2016 *Scientific Rep.* Forests as a natural seismic metamaterial: Rayleigh wave bandgaps induced by local resonances, **6** 19238.
[15] Aravantinoszafiris N and Sigalas M M 2015 *J. Appl. Phys.* Large scale phononic metamaterials for seismic isolation, **118** 064901.
[16] Shi X, Shu H, Zhou H, et al. 2017 *J. Appl. Phys.* SH wave propagation in joined half-spaces composed of elastic metamaterials, **122(21)** 215104.
[17] Geers M G D, Kouznetsova V G, Massart T J, et al. 2009 Computational homogenization of structures and materials.
[18] Pham K, Kouznetsova V G and Geers M G D 2013 *J. Mech. Phys. Solids* Transient computational homogenization for heterogeneous materials under dynamic excitation, **61(11)** 2125-46.
[19] Wu Y, Lai Y and Zhang Z Q 2007 *Phys. Rev. B Condensed Matt.* Effective medium theory for elastic metamaterials in two dimensions, **76(20)** 205313
[20] Roca D, Lloberas-Valls O, Cante J, et al. 2018 *Comput. Methods Appl. Mech. Eng.* A computational multiscale homogenization framework accounting for inertial effects: Application to acoustic metamaterials modelling, **330** 415-46.
[21] Huang H H, Sun C T and Huang G L 2009 *Int. J. Eng. Sci.* On the negative effective mass density in acoustic metamaterials, **47(4)** 610-7.
[22] Sridhar A, Kouznetsova V G and Geers M G D 2016 *Computation. Mech.* Homogenization of locally resonant acoustic metamaterials towards an emergent enriched continuum, **57(3)** 423-35.
[23] Torrent D and Sánchez-Dehesa J 2009 *Phys. Rev. Lett.* Radial wave crystals: radially periodic structures from anisotropic metamaterials for engineering acoustic or electromagnetic waves, **103(6)** 064301.
[24] Shu H, Xu L, Shi X, et al. 2016 *J. Appl. Phys.* Traveling Lamb wave in elastic metamaterial layer, **120(16)** 165103.
[25] Williams E G, Roux P, Rupin M, et al. 2015 *Physic. Rev. B* Theory of multiresonant metamaterials for A0 Lamb waves, **91(10)** 104307.
[26] Zhu R, Liu X N, Huang G L, et al. 2012 *Physic. Rev. B* Microstructural design and experimental validation of elastic metamaterial plates with anisotropic mass density, **25(14)** 144307.
[27] Krödel S, Thomé N and Daraio C 2015 *Extreme Mech. Lett.* Wide band-gap seismic metastructures, **4** 111-7.
[28] Ping S 1995 *Waves Random Complex Media* Introduction to Wave Scattering, Localization and Mesoscopic Phenomena, **17(2)** 235-7.
[29] Srivastava A 2015 *Int. J. Smart Nano Mater.* Elastic Metamaterials and Dynamic Homogenization: A Review, **6(1)** 41-60.
[30] Shen X H, Sun C T, Barnhart M V and Huang G L 2018 *J. Vib. Acoust.* Analysis of Dynamic Behavior of the Finite Elastic Metamaterial-Based Structure with Frequency Dependent Properties, **140(3)** 031012.
[31] Achenbach J D 1980 Wave Propagation in Elastic Solids, North-Holland.

Acknowledgments
This work was funded by the project (grant number 51875112) supported by the National Natural Science Foundation of China, the project (grant number LBH-Q15029) supported by the Postdoctoral Scientific Research Developmental Fund of Heilongjiang Province of China.

IOP Conf. Series: Materials Science and Engineering **649** (2019) 012010 doi:10.1088/1757-899X/649/1/012010

Using Absorbent Paper Strips for the Collection of Cell-Free DNA in Patients with Periodontal Diseases

Boonyanit Thaweboon[1], Sukhumchawee Suwannagindra[2], Varunee Kerdvongbundit [2] and Sroisiri Thaweboon[1]

[1]Department of Oral Microbiology

[2]Department of Oral Medicine and Periodontology, Faculty of Dentistry, Mahidol University, Bangkok 10400, Thailand

boonyanit.tha@mahidol.ac.th, khumwee@gmail.com, vasria@hotmail.com, sroisiri.tha@mahidol.ac.th

Abstract. The cell-free DNA in gingival crevicular fluid (GCF) is a recent interesting diagnostic marker in patients with periodontal disease. The methods used to collect GCF are very important to get the proper specimen. There are several techniques used for sample collection, such as gingival washing or using a capillary tube, however, these methods are not practical to perform in the clinic since they need well-trained examiner. This study aimed to use absorbent paper strips to collect cell-free DNA fragment in GCF of patients as a diagnostic measure for periodontal diseases comparing to conventional washing technique. Thirty-nine periodontitis and 26 gingivitis teeth were selected from 26 patients according to clinical parameters. The GCF sample was randomly collected from each tooth by either paper strips or washing technique. Then, after one week, the collection of samples from the same tooth was repeated with another technique. The samples were centrifuged to get cell-free DNA in the supernatant and extracted by InstaGene Matrix. The concentration and purity of extracted DNA were determined by NanoDrop spectrophotometer. Three sets of specific primers to the human β-globin gene were used to evaluate the DNA fragment lengths, by amplifying 110 base pair (bp), 536 bp, and 2000 bp products in the polymerase chain reaction (PCR). The results showed no significant differences in the cell-free DNA fragment concentrations and the prevalence of PCR products between using absorbent paper strips and washing technique in either gingivitis or periodontitis groups. Therefore, with the ease and rapidity of the technique, using the absorbent paper strips for the quantitative and qualitative investigation of cell-free DNA fragment in patients with periodontal diseases instead of washing technique is recommended. Further studies are still required to confirm this finding in a large group of patients as well as in patients with other systemic health problems.

1. Introduction

Periodontal diseases are common oral infectious diseases found in people worldwide. The diseases are characterized by an inflammation of periodontal tissue or supporting tissue of the teeth (gingiva, bone, and periodontal ligament) [1]. They are the major causes of tooth loss in an adult that have affected to function of chewing, talking, smiling and social appearance. Gingivitis is the initial stage of periodontal diseases which is a reversible condition and can usually be eliminated by professional prophylaxis. In contrast, the advanced form is called periodontitis of which the patients' tissues undergo irreversible destruction during the inflammatory process.

Gingival crevicular fluid (GCF) is regarded as a promising mean in investigations associated with the diagnosis, pathogenesis, and prognosis of periodontal diseases. The amount of GCF and flow rate is an indicator of changes in vascular permeability which depends on the degree of inflammation. GCF is composed of inflammatory cells and mediators, bacteria, enzymatic and non-enzymatic components as well as cell-free DNA fragments from the damaged cells which are biomarkers for the diagnosis of periodontal diseases [2-5]. It can be collected from the gingival sulcus or periodontal pocket of the patients with periodontal diseases. Thaweboon *et al* has proposed using different polymerase chain reaction (PCR) products of cell-free DNA i.e. 110 base pairs (bp), 536 bp, and 2000 bp, specific to the human β-globin gene in GCF to classify the periodontal conditions of the patients [6] The 2000 bp and 536 bp products were found mostly from periodontitis teeth while only the 536 bp product was from gingivitis teeth.

There are several techniques used for GCF collection, such as gingival washing technique and the use of capillary tube [7]. However, these methods are not feasible to perform in the clinic since they need well-practiced examiner. In the gingival washing method, an isotonic solution such as a phosphate buffer solution (PBS) is pipetted to the gingival sulcus then all the contents and fluid from the sulcus will be aspirated back. This method is highly sensitive but requires the participation of a trained, experienced investigator to collect samples. For the capillary method, the small capillary tube is placed at the entrance of gingival sulcus until an adequate amount of fluid is obtained. Even though this method seems to be simple but it depends on crucial skill and takes time especially in healthy teeth. Using absorbent paper to collect GCF can be an alternative, especially for the detection of cell-free DNA fragment from the patients since it is fast, easy and can be applied to a particular site.

The aim of this study was to use absorbent paper strips to collect cell-free DNA fragment in GCF of patients as a diagnostic measure for periodontal diseases compared to conventional washing technique.

2. Materials and Methods

The study was conducted in the Oral Medicine and Periodontology Clinic, Faculty of Dentistry, Mahidol University. Twenty-six patients, who had at least 4 chronic periodontitis teeth, were recruited in this study. Patients who had any history of smoking, a systemic disease affecting periodontal disease progression (e.g. diabetes, immune deficiency disease), pregnancy or lactation at the time of the study, and had taken the antibiotic medication within 6 weeks prior to the study were excluded. The study protocol was approved by the committee of Institutional Review Board of the Faculty of Dentistry/Faculty of Pharmacy, Mahidol University, Thailand (MU-DT/PY-IRB 2013/033.1009).

In the 1st visit, clinical parameter measurements were performed on each patient. The GCF sample was randomly collected from the selected tooth by washing technique or using an absorbent paper strip in the 2nd visit. Then, another technique was done during the 3rd visit. The time interval between visits was 7 to 10 days.

2.1. GCF Collection

Twenty-six patients were recruited in the study (13 men and 13 women, aged 53.10$\pm$5.28 years). One gingivitis tooth that was not adjacent to the periodontitis tooth was selected for the gingivitis group and one or two periodontitis teeth in the different quadrant from each patient were selected for the periodontitis group.

Modified washing technique [6]: a 5 μL aliquot of sterile phosphate buffered saline (PBS) solution (PBS, pH 7.2) was used to gently flush at the entrance of the pocket or sulcus by micropipette. Afterward, the aspirated contents and fluid were collected in a 1.5 ml sterile plastic microtube. The procedure was repeated 5 times at buccal aspect and another 5 times at lingual aspect of the selected teeth.

Absorbent paper strip technique: the top three depths of pocket from each tooth were selected for this collection. An absorbent paper strip was gently placed into each selected site until light resistance is felt.

It was left for 30 seconds [8]. All 3 paper strips were put in a 1.5 ml microtube containing 100 µl PBS at 4°C. The samples were eluted from the paper strips by a vortex mixer for 1 minute.

To separate the host cells, the samples from both techniques were centrifuged at 3,000 g for 10 minutes at 4°C. The supernatant, a cell-free fraction, was kept in 80% of ethanol at -20°C until DNA extraction was performed.

2.2. DNA Extraction and Polymerase Chain Reaction (*PCR*)

The cell-free fraction in ethanol was centrifuged at 13,000 g for 10 minutes at 4°C. The supernatant was removed and the tube was placed upside-down on a paper towel to remove excess ethanol from the pellet. The dried pellet was dissolved in 30 µL of sterile deionized water and mixed with 200 µL of DNA extraction liquid (InstaGene Matrix; Bio-Rad, Hercules, CA, USA) according to the manufacturing protocol. The concentration and purity of DNA sample were measured by NanoDrop2000c spectrophotometer (Thermo Scientific, USA).

Conventional PCR using 3-primer set specific to the human β-globin gene was done to amplify PCR products (110 bp, 536 bp, and 2000 bp) using forward and reverse primers as mentioned previously by Doan et al [6, 9]. DNA extracted from buccal cells of a healthy volunteer and deionized water was used as positive and negative controls, respectively. Afterward, all PCR products were observed on 1.5% agarose gel electrophoresis and ethidium bromide stain.

2.3. Statistical Analysis

DNA concentration from the two different techniques in the gingivitis or periodontitis groups were compared and analyzed by Pair T-test. Comparison of the positive number of each PCR product between washing technique and paper strips technique was evaluated by McNemar's test. The significance level was set at a p-value less than 0.05.

3. Results

Cell-free DNA concentrations obtained by the paper strips and washing techniques in gingivitis or periodontitis groups were not significant differences between the two techniques (table 1). In addition, using paper strips seemed to yield a little more concentration of cell-free DNA than washing technique in the gingivitis group. Regarding the PCR product yield, there were no significances between the two techniques (table 2).

Table 1. Cell-free DNA concentrations between paper strips and washing techniques in gingivitis and periodontitis groups.

	Concentration of Cell-Free DNA (ng/µl)		
	Paper strips	Washing technique	p-value
Gingivitis (N=26)	10.44±7.61	8.81+7.76	>0.05
Periodontitis (N=39)	17.83+14.21	20.50±15.93	>0.05

Table 2. The number of samples with positive PCR products in the gingivitis and periodontitis groups.

	Number of samples with positive PCR products			
	PCR Product (bp)	Paper strips	Washing technique	p-value
Gingivitis (N=26)	536	19/26	18/26	> 0.05
	20000	0/26	0/26	> 0.05
Periodontitis (N=39)	536	35/39	39/39	> 0.05
	2000	30/39	31/39	> 0.05

4. Discussion

The cell-free DNA fragment found in serum, plasma, or other body fluids is a current interesting diagnostic tool in various medical fields [10-12] and dentistry [6, 9]. The cell-free DNA fragment released by dead cells in the fluid can directly represent the severity of inflammation and cell destruction (17, 18) as appeared in periodontal diseases. The present study showed that both techniques were suitable to collect the human cell-free DNA fragment in GCF because all samples could be amplified the 110 bp PCR product of the human β-globin gene. Although no significant difference in cell-free DNA concentration was noticed between the two techniques, the paper strips seemed to provide a higher yield, especially in the gingivitis group. This might be from the fact that paper strips could easily absorb all GCF in the shallow pocket of gingivitis condition while the washing technique was quite difficult to aspirate all the fluid back since it usually leaked out from the pocket. On the other hand, with the presence of deep pocket in the periodontitis group, the flushing and aspirating by washing technique could be done without any escape of the fluid.

With regard to the quality of cell-free DNA, the present study has shown the consistent result to that of our previous study [6] which stated that the 2000 bp PCR product could be as a biomarker to differentiate the periodontitis from gingivitis conditions. Furthermore, there was no significant difference in the prevalence of 536 bp and 2000 bp PCR products between using paper strips and washing technique. Several studies described the benefit of using absorbent paper strips for the quantitation of various inflammatory cytokines (IL-2,-4,-6,-8,-10, TNF-α,), enzymes (MMP-1,-2,-8,-13, myeloperoxidase, elastin) and proteins (osteocalcin, calprotectin, melatonin) in GCF to support clinical periodontal diagnosis in both pediatric and adult patients [7, 8, 13-16]. Therefore, it might imply that paper strips are efficiently used to collect GCF for the detection of cell-free DNA fragments which can be adequately recovered from the paper for further disease diagnosis in patients with periodontal diseases equivalent to the conventional washing technique.

5. Summary

With regard to the quantity and quality of cell-free DNA fragment, the absorbent paper strips are useful means for GCF collection in patients with periodontitis instead of the washing technique.

References

[1] Kinane D F, Stathopoulou P G and Papapanou P N 2017 *Nat. Rev. Dis. Primers* Periodontal diseases, **3** 17038.

[2] Kumar A K, Reddy N R, Babu M, Kumar P M, Reddy V S and Chavan C V 2013 *Contemp. Clin. Dent.* Estimation of prostaglandin E2 levels in gingival crevicular fluid in periodontal health, disease and after treatment, **4** 303-6.

[3] Loos B G 2005 *Periodontol 2000* Host-derived diagnostic marker for periodontitis: Do they exist in gingival crevice fluid?, **39** 53-72.

[4] Stadler A F, Angst P D, Arce R M, Gomes S C, Oppermann R V and Susin C 2016 *J. Clin. Periodontol.* Gingival crevicular fluid levels of cytokines/chemokines in chronic periodontitis: a meta analysis, **43** 727-45.

[5] Teles R, Sakellari D, Teles F, Konstantinidis A, Kent R and Socransky S, et al. 2010 *J. Periodontol.* Relationships among gingival crevicular fluid biomarkers, clinical parameters of periodontal disease, and the subgingival microbiota, **81** 89-98.

[6] Thaweboon B, Laohapand P, Amornchat C, Matsuyama J, Sato T and Nunez P P, et al. 2010 *J. Periodontal. Res.* Host beta-globin gene fragments in crevicular fluid as a biomarker in periodontal health and disease, **45** 38-44.

[7] Majeed Z N, Phillip K, Alabsi A M, Pushparajan S and Swaminathan D 2016 *Dis. Markers* Identification of gingival crevicular fluid sampling, analytical methods, and oral biomarkers for the diagnosis and monitoring of periodontal diseases: A systematic review, **2016** 1804727.

[8] Wassall R R and Preshaw P M 2016 *Periodontol 2000* Clinical and technical considerations in the analysis of gingival crevicular fluid, **70** 65-79.

[9] Doan T M, Kriangcherdsak Y, Thaweboon S and Thaweboon B 2015 *Int. J. Oral. Maxillofac. Implants.* Evaluation of host beta-globin gene fragment lengths in peri-implant crevicular fluid during the wound healing process: a pilot study, **30** 1295-302.

[10] Gögenur M, Burcharth J and Gögenur I 2017 *Crit. Care.* The role of total cell-free DNA in predicting outcomes among trauma patients in the intensive care unit: a systematic review, **21** 14.

[11] Urbanova M, Plzak J, Strnad H and Betka J 2010 *Cell. Mol. Biol. Lett.* Circulating nucleic acids as a new diagnostic tool, **15** 242-59.

[12] Frank M O 2016 *Biol. Res. Nus.* Circulating cell-free DNA differentiates severity of inflammation, **18** 477-88.

[13] Alfant B, Shaddox L M, Tobler J, Magnusson I, Aukhil I and Walker B C 2008 *J. Periodontol* Matrix metalloproteinase in children with aggressive periodontitis, **79** 819-26.

[14] Bullon P, Goberna B, Guerrero J M, Segura J J, PerezCano R, and Martinez-Sahuquillo A 2005 *J. Periodontol.* Serum, saliva, and gingival crevicular fluid osteocalcin: Their relation to periodontal status and bone mineral density in postmenopausal women, **76** 513-9.

[15] Emingil G, Gurkan A, Atilla G and Kantarci A 2011 *J. Periodontol.* Subantimicrobial-dose doxycycline and cytokine-chemokine levels in gingival crevicular fluid, **82** 452-61.

[16] Kaner D, Bernimoulin J P, Kleber B M, Heizmann W R and Friedmann A 2006 *J. Periodontal. Res.* Gingival crevicular fluid levels of calprotectin and myeloperoxidase during therapy for generalized aggressive periodontitis, **41** 132-9.

Acknowledgments

This study was supported by grants number 02/2554 from the Faculty of Dentistry, Mahidol University, Bangkok, Thailand. The authors appreciatively acknowledge Associate Professor Penpan Laohapand for her valuable comments and encouragement during the study.

IOP Conf. Series: Materials Science and Engineering **649** (2019) 012011 doi:10.1088/1757-899X/649/1/012011

The Natural Characteristics and Forced Response of O-shaped Plate Under Isotropic and Orthotropic Materials

Dongyan Shi[1, 4], Wenhui Ren[1], Zhongxin Lan[2] and Yongxin Li[3, 4]

[1]College of Mechanical and Electrical Engineering, Harbin Engineering University, 150001, Harbin, China

[2]College of Automation, Harbin Engineering University, 150001, Harbin, China

[3]School of mechanical engineering, Yanshan University, 066004, Qinhuangdao, China

[4]Heavy-duty Intelligent Manufacturing Equipment Innovation Center of Hebei Province, 066004, Qinhuangdao, China

shidongyan@hrbeu.edu.cn, renwenhui1@outlook.com, 2777474931@qq.com, liyongxin@ysu.edu.cn

Abstract. In this paper, the O-shaped plate with different materials are used to analyze their natural frequencies at different boundaries and different plate thicknesses and the forced response under the action of point forces. By the improved Fourier series method, the discontinuity of coupling system at all boundaries in the overall solution domain is overcame. Through the energy principle and the Rayleigh-Ritz technology, the matrix equation of the system governing equations can be obtained. In addition, based on various materials, boundary conditions and thicknesses, some new results have been demonstrated, which may become the foundation for the future research.

1. Introduction

In recent years, O-shaped plate structure is widely used in real life. It is of great significance to study the natural characteristics and forced response of the O-shaped plate structure.

Some literatures about the coupled plate structures are listed here. Shen et al. [1] studied combination of rectangular thin isotropic plates. Li [2] used a scaling approach to analyzing the line-coupled plates at high frequencies. Tian et al. [3] introduced the vibration response of an L-shaped plate under a point force or a moment excitation. Du et al. [4] applied an analytical method to determin the vibrations of two plates which are generally supported along the boundary edges, and elastically coupled together at an arbitrary angle. Boisson et al. [5] discussed the relationship between factors and the vibration energy transmission characteristics. The above articles investigate different tools to consider the vibration characteristics. But the existing theoretical analysis of O-shaped coupled plate considers the bending moment between the plates, while ignoring the lateral shear, longitudinal and in-plane shearing effects in the plate. Studies have shown large deviations occur in the displacement response of the coupled plate structure using the modeling technique that does not include the in-plane vibration component. Therefore, it is very important to establish a more general vibration model.

In view of the above defects, the displacement admissible function is optimized based on the Fourier series method. At the end, the natural characteristics and forced response of the O-shaped coupling plate are discussed.

2. Theoretical Formulations

2.1. Description of the Model

As shown in figure 1, the O-shaped plate system is established to analyze the natural characteristics and forced response. The geometrical dimensions of the O-shaped plate system and the coordinate system used in this paper have been shown in figure 1. In this paper, the plate dimensions are $a_1=a_2=a_3=a_4=b=1$m. In table 1, we list four types of material properties.

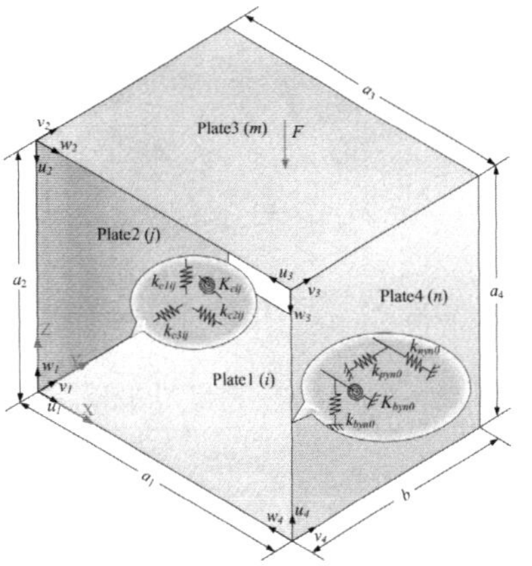

Figure 1. Analytical model.

Table 1. Four different types of material properties.

	Material I	Material II	Material III	Material IV
Material	Steel	Aluminum	Graphite /epoxy	Glass /epoxy
E_1[GPa]	216	71	185	39
E_2[GPa]	216	71	10.5	8.4
G_{12}[GPa]			7.3	4.2
μ	0.3	0.3	0.28	0.26
ρ[kg/m^3]	7800	2700	1600	1600

2.2. Energy Expressions of the O-shaped Plate System

In this paper, the energy equation will be used to describe the motion equation of the O-shaped plate system. The Lagrange equations of the system shown in figure 1 can be expressed as:

$$L_{\text{plane}}=(U_{\text{jbending}}+U_{\text{jin-plane}}+U_{\text{coupling}})-(T_{\text{jbending}}+T_{\text{jin-plane}})-W_{\text{plane}} \tag{1}$$

where U_{jbending} (j=1,2,3,4) and $U_{\text{jin-plane}}$ represent respectively the total potential energy associated with bending vibration and the total potential energy associated with in-plane vibration in the j'th plate; T_{jbending} and $T_{\text{jin-plane}}$ represent respectively the correspondence of the j'th plate kinetic energy of bending and in-plane vibration; U_{coupling} indicates the elastic potential energy stored in the coupling boundary spring. Take plate 1 as example, potential energy and kinetic energy can be expressed as:

$$U_{\text{1bending}}=\frac{D_1}{2}\int_0^{a_1}\int_0^{b}\left\{\left(\frac{\partial^2 w_1}{\partial x_1^2}\right)^2+\left(\frac{\partial^2 w_1}{\partial y_1^2}\right)^2+2\mu_1\frac{\partial^2 w_1}{\partial x_1^2}\frac{\partial^2 w_1}{\partial y_1^2}+2(1-\mu_1)\left(\frac{\partial^2 w_1}{\partial x_1\partial y_1}\right)^2\right\}dx_1dy_1+\frac{1}{2}\int_0^{b}\left[k_{bx10}w_1^2+K_{bx10}\left(\frac{\partial w_1}{\partial x_1}\right)^2\right]_{x_1=0}dy_1 \tag{2}$$

$$+\frac{1}{2}\int_0^{b}\left[k_{bx11}w_1^2+K_{bx11}\left(\frac{\partial w_1}{\partial x_1}\right)^2\right]_{x_1=a_1}dy_1+\frac{1}{2}\int_0^{a_1}\left[k_{by10}w_1^2+K_{by10}\left(\frac{\partial w_1}{\partial y_1}\right)^2\right]_{y_1=0}dx_1+\frac{1}{2}\int_0^{a_1}\left[k_{by11}w_1^2+K_{by11}\left(\frac{\partial w_1}{\partial y_1}\right)^2\right]_{y_1=b_1}dx_1$$

$$U_{\text{1in-plane}}=\frac{G_1}{2}\int_0^{a_1}\int_0^{b}\left\{\left(\frac{\partial u_1}{\partial x_1}+\frac{\partial v_1}{\partial y_1}\right)^2-2(1-\mu_1)\frac{\partial u_1}{\partial x_1}\frac{\partial v_1}{\partial y_1}+\frac{1-\mu_1}{2}\left(\frac{\partial v_1}{\partial x_1}+\frac{\partial u_1}{\partial y_1}\right)^2\right\}dx_1y_1+\frac{1}{2}\int_0^{b}\left[k_{nx10}u_1^2+k_{px10}v_1^2\right]_{x_1=0}dy_1 \tag{3}$$

$$+\frac{1}{2}\int_0^{b}\left[k_{nx11}u_1^2+k_{px11}v_1^2\right]_{x_1=a_1}dy_1+\frac{1}{2}\int_0^{a_1}\left[k_{ny10}v_1^2+k_{py10}u_1^2\right]_{y_1=0}dx_1+\frac{1}{2}\int_0^{a_1}\left[k_{ny11}v_1^2+k_{py11}u_1^2\right]_{y_1=b_1}dx_1$$

$$U_{\text{coupling12}} = \frac{1}{2}\int_0^b \Big[K_{c12}\left(\partial w_1/\partial x_1\big|_{x_1=0} - \partial w_2/\partial x_2\big|_{x_2=a_2}\right)^2 + k_{c112}\left(w_1\big|_{x_1=0} - w_2\big|_{x_2=a_2}\cos\theta + u_2\big|_{x_2=a_2}\sin\theta\right)^2$$
$$+ k_{c212}\left(u_1\big|_{x_1=0} - u_2\big|_{x_2=a_2}\cos\theta - w_2\big|_{x_2=a_2}\sin\theta\right)^2 + k_{c312}\left(v_1\big|_{x_1=0} - v_2\big|_{x_2=a_2}\right)^2\Big]dy \quad (4)$$

$$U_{\text{coupling41}} = \frac{1}{2}\int_0^b \Big[K_{c41}\left(\partial w_4/\partial x_4\big|_{x_4=0} - \partial w_1/\partial x_1\big|_{x_1=a_1}\right)^2 + k_{c141}\left(w_4\big|_{x_4=0} - w_1\big|_{x_1=a_1}\cos\theta + u_1\big|_{x_1=a_1}\sin\theta\right)^2$$
$$+ k_{c241}\left(u_4\big|_{x_4=0} - u_1\big|_{x_1=a_1}\cos\theta - w_1\big|_{x_1=a_1}\sin\theta\right)^2 + k_{c341}\left(v_4\big|_{x_4=0} - v_1\big|_{x_1=a_1}\right)^2\Big]dy \quad (5)$$

where k_{bx10}, k_{bx11}, k_{by10}, k_{by11}, k_{nx10}, k_{nx11}, k_{ny10}, k_{ny11}, k_{px10}, k_{px11}, k_{py10}, k_{py11}, k_{c1}, k_{c2} and k_{c3} are the linear springs and K_{bx10}, K_{bx11}, K_{by10}, K_{by11} and K_c are the rotational springs. ρ and μ represent mass density and Poisson's ratio. ω is the circle frequency. D and G represent the bending stiffness and tensile stiffness of plate structure. θ is the coupling angle between the two plates.

$$T_{1\text{bending}} + T_{1\text{in-plane}} = \frac{1}{2}\rho_1 h_1 \omega^2 \int_0^{a_1}\int_0^b (w_1^2 + u_1^2 + v_1^2)dx_1 dy_1 \quad (6)$$

W_{panel} is the work done by the external stimulus on the structure. It is easy to get that $W_{\text{panel}}=Fw(x_e,y_e)$. Subscript 1 is used to indicate all these quantities for board 1.

2.3. Admissible Displacement Pressure Functions

For the coupled-plate structure as shown in the figure, when the angle $\theta \neq 0$ is considered, there is a coupling phenomenon. Thus, the plate structure needs to consider two kinds of vibration components at the same time. Taking plate 1 as example, the bending vibration displacement w_1 and in-plane vibration displacement u_1 and v_1 can be expressed as:

$$\tau_1(x_1,y_1) = \sum_{m=0}^{\infty}\sum_{n=0}^{\infty}\Gamma_{1mn}\cos\lambda_{a_1m}x_1\cos\lambda_{bn}y_1 + \sum_{m=-4}^{-1}\sum_{n=0}^{+\infty}\Gamma_{1mn}^1\sin(\lambda_{a_1m}x_1)\cos(\lambda_{bn}y_1) + \sum_{m=0}^{+\infty}\sum_{n=-4}^{-1}\Gamma_{1mn}^2\cos(\lambda_{a_1m}x_1)\sin(\lambda_{bn}y_1) \quad (7)$$

in which λ_{a1m}=mπ/a_1, λ_{bn}=nπ/b. These functions could eliminate the discontinuous or jumping phenomenon in the boundaries. Substituting equations (1)-(7) into equation (8) according to the Rayleigh-Ritz technology, the equation (9) can be obtained and expressed in a matrix form:

$$\frac{\partial L_{\text{panel}}}{\partial \Gamma_{jmn}} = \frac{\partial U_{\text{panel}}}{\partial \Gamma_{jmn}} - \frac{\partial T_{\text{panel}}}{\partial \Gamma_{jmn}} - \frac{\partial W_{\text{panel}}}{\partial \Gamma_{jmn}} = \frac{\partial W_{\text{ext}}}{\partial \Gamma_{jmn}} \quad (8)$$

$$\left[K_p - \omega^2 M_p\right]\Gamma = F \quad (9)$$

where K_p, M_p and F are the stiffness, mass and force matrices of the plate. Γ is the unknown Fourier coefficients vectors of the plate displacement.

3. Numerical Calculation and Analysis

3.1. Study on the Convergence Analysis

Because the Fourier series method is applied to the O-shaped plate system, it is necessary to study its convergence and accuracy. In table 2, the material was selected as material I, and the thickness of the plate was 8 mm. The boundary conditions of table 2 are that both the coupled edge and the uncoupled edge are fixed.

Table 2. Convergence and accuracy of the first eight frequency f [Hz] for the O-shaped plate.

$M_p \times N_p$	Mode Number							
	1	2	3	4	5	6	7	8
8×8	72.50	147.87	148.37	218.81	265.10	266.35	332.46	333.57
9×9	72.50	147.87	148.13	218.44	265.10	266.36	332.44	333.06
10×10	72.50	147.87	148.13	218.44	265.10	266.36	332.44	333.05
11×11	72.50	147.87	148.02	218.27	265.10	266.36	332.44	332.81
FEM	72.92	148.70	148.70	219.11	266.57	267.84	333.98	333.98

Under the premise of satisfying the correctness of the results and calculating efficiency, in the following numerical examples, all the truncated values of the coupled system are Mp=Np=9. At the end of every table, the results obtained by finite element method (FEM) have also been given, which shows good agreement with the results obtained by the present method.

3.2. The Natural Characteristics of the O-shaped Plate System

In order to deepen the intuitive understanding of the vibration characteristics of O-shaped plate system, the displacement mode shapes of the O-shaped plate are given in figure 2. The material and dimensions are the same as those in table 2. The boundary condition of the coupling edge of coupling plate is the fixed support, and the remaining boundary conditions are free, simple and fixed, respectively, which are represented by the letters a, b, and c in figure 2. From the data given in the figure 2, we can know that the results of this paper are consistent with the results of finite element simulations. The more boundary conditions that are fixed, the higher the natural frequency of the O-shaped plate system. From the mode shapes, we can directly see the displacement distribution of the O-shaped plate at the natural frequencies. The mode shapes provide the basis for the analysis of the vibration characteristics of the structural system.

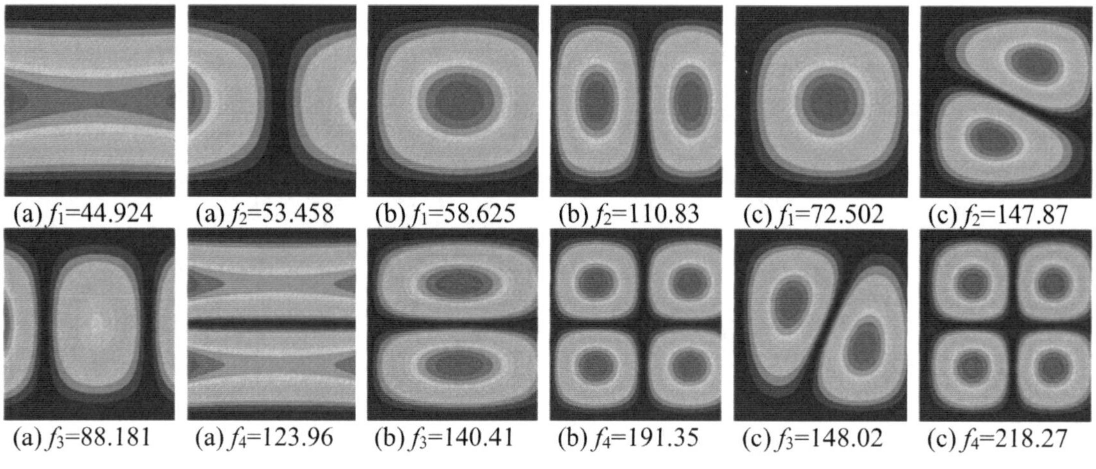

Figure 2. The mode shapes and natural frequency f [Hz] of the O-shaped plate system.

As shown in figure 3, we analyze effects of different thickness ratios. The A, B, C and D in the serial number represents plate thickness of 1, 2, 4, 8mm, respectively. The material, boundary condition and dimensions are the same as those in table 2. It can be concluded from the table that when the thickness of the four plates is the same, the natural frequency increases as the thickness of the plate increases. When the thicknesses of the four plates are not the same, as the thickness of the plate increases, the natural frequency increases but the increase at low frequencies is not very noticeable.

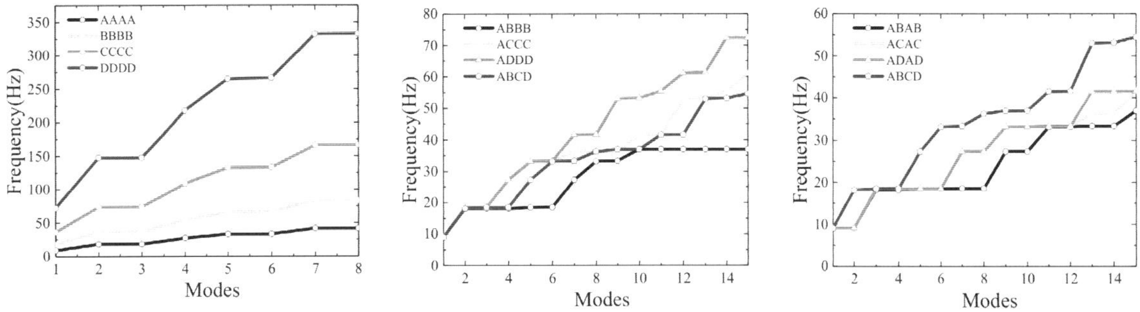

Figure 3. The natural frequency *f* [Hz] of the O-shaped plate system under different plate thickness.

3.3. Response of the O-shaped Plate System

This section focuses on the influence of forces on the response of the O-shaped plate system. The geometric dimensions are the same as in figure 1. The plates are subjected to fixed boundary conditions. The point forces are applied on every plate at (0.5,0.5) whose magnitude is *F*=1N. Figure 4 has shown the displacement response of O-shaped plate under different materials in the frequency range of 0-500 Hz. The first and third curves are the result of the present method. the second and fourth curves are the results of FEM. The results obtained by the present method and FEM match very well, which proves the correctness of the proposed method in predicting the vibration response of the O-shaped plate system. Form the response problem of the plate, for isotropic materials, it can be seen from the response curve that when the elastic modulus of the plate material and the density are smaller, the displacement of the plate is rather large. For orthotropic materials, the displacement of plates is different due to the different natural frequencies.

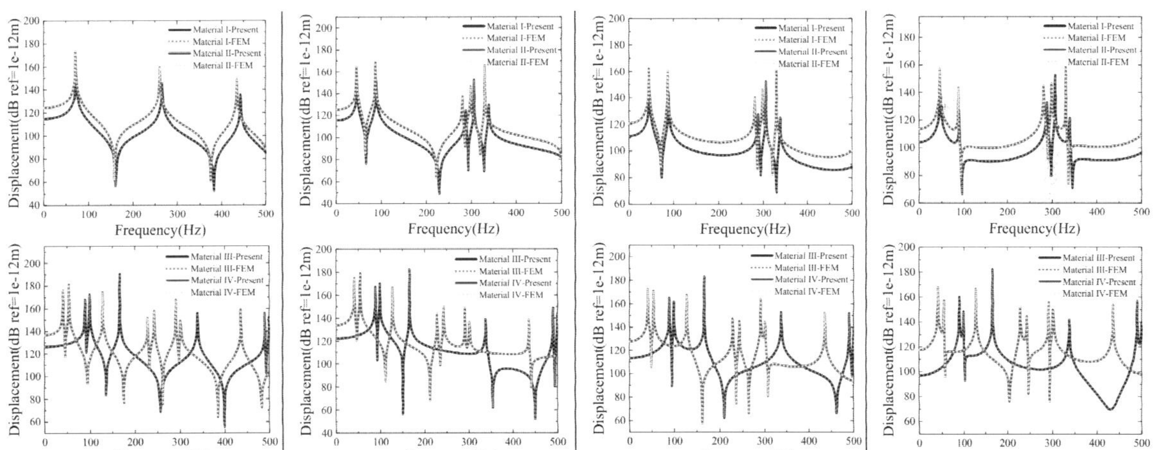

Displacement at (0.5,0.5) in plate 1 | Displacement at (0.4,0.4) in plate 2 | Displacement at (0.3,0.3) in plate 3 | Displacement at (0.2,0.2) in plate 4

Figure 4. Responses of O-shaped plate system under different material and excitation of point forces.

4. Summary

In this paper, Fourier series method is applied to study the characteristics of O-shaped plate system. From the results of this paper, it can be seen the method of this paper is correct. The natural frequency will increase as the increase of the spring stiffness value. For the response problem of the isotropic plate, it can be seen from the response curve that when the elastic modulus of the plate material and the density are smaller, the displacement of the plate is rather large. For orthotropic materials, the displacement of the plates is different due to the different natural frequencies.

References

[1] Shen Y and Gibbs B M 1986 *J. sound vib.* An approximate solution for the bending vibrations of a combination of rectangular thin plates, **105(1)** 73-90.

[2] Li X 2013 *J. Sound Vib.* A scaling approach for high-frequency vibration analysis of line-coupled plates, **332(18)** 4054-8.

[3] Tian R and Andy C C, et al. 2011 *J. Sound Vib.* Vibration of L-shaped plates under a deterministic force or moment excitation: a case of statistical energy analysis application, **330(20)** 4780-97.

[4] Du J and Li W, et al. 2011 *J. Sound Vib.* Free vibration of two elastically coupled rectangular plates with uniform elastic boundary restraints, **330(4)** 788-804.

[5] Boisson C, et al. 1982 *J. Sound Vib.* Energy transmission in finite coupled plates, part II: Application to an L shaped structure, **81(1)** 93-105.

Acknowledgements

This paper is funded by the International Exchange Program of Harbin Engineering University for Innovation-oriented Talents Cultivation. The works also supported by the National Natural Science Foundation of China (No.51679056) and Natural Science Foundation of Heilongjiang Province of China (E2016024).

Research on Artificial Post-Treatment Technology of FDM Forming Parts

Jinjin Liu[1, 2], Hai Gu[2, 3], Bin Li[2, 3], Lu Zhu[3], Jie Jiang[2, 3] and Jie Zhang[2, 3]

[1]Nantong Institute of Technology of 3D Technology Co. Ltd,226002 Nantong Jiangsu, China

[2]Jiangsu Key Laboratory of 3D Printing Equipment and Application Technology,226002 Nantong Jiangsu, China

[3]School of Mechanical Engineering, Nantong Institute of Technology,226002 Nantong Jiangsu, China

970478720@qq.com, guhaint@ntit.edu.cn, libin19@ntit.edu.cn, 1974154843@qq.com, jiangjie@ntit.edu.cn, zjzcc_2008@yeah.net

Abstract. 3D printing technology is a high and new technology gradually developed from the mid-1980s, leading the manufacturing technology of the third revolution. The 3D printing process includes modeling, data processing, selective process printing and post-processing. FDM technology is the most widely used technology at present, but the molding accuracy is not high. Due to the "step effect", the printed parts have rough surface, obvious stripes, poor surface quality, and cannot meet the customer's or specified requirements, so post-processing is very important. This paper mainly studies and summarizes the manual post-processing technology of FDM printed parts, and provides the specific implementation method of post-processing, providing reference for the post-processing of FDM formed parts and other forming processes.

1. Introduction

3D printing technology is a kind of rapid prototyping technology, which appeared in the late 1980s [1]. It is a technology of solid model manufacturing by means of layer-by-layer accumulation of CAD design data materials. Compared with traditional reduction manufacturing, 3D printing technology is a "bottom-up" additive manufacturing. The 3D printing process includes data processing, forming and post-processing. The data processing stage is the basis for obtaining a good model, while the post-processing stage reflects the value of the model. After the printing model is finished, the formed parts are taken out from the forming platform, and then to remove support, sand, polish, secondary cute, or in the hign-temperature furnace for sintering according to different processes, so as to further improve the surface quality and strength.

2. Molten Deposition Manufacturing

2.1. Principle of Molten Deposition Manufacturing

The principle of FDM technology is shown in figure 1. The filaments of hot-melt materials are heated and melted, driven by the operation of gears, and then they are uniformly extruded from the nozzle at the bottom of the nozzle. During the printing work, the nozzle moves along the X axis and the working platform moves along the Y axis and Z axis. After the extrusion is bonded to the upper layer of material, and then the printing platform will drop one layer thick and repeat printing until it is completed [2].

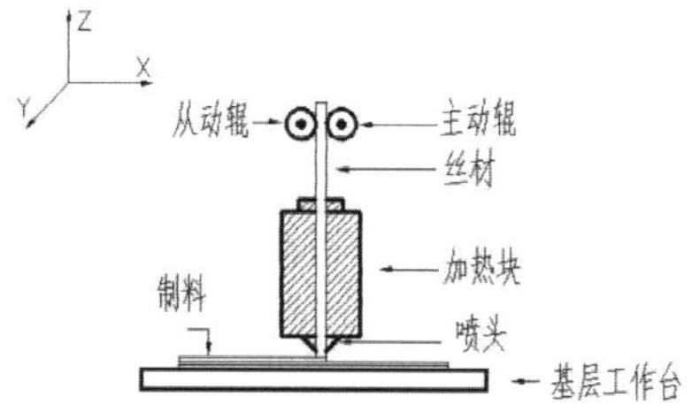

Figure 1. FDM principle diagram.

2.2. FDM Forming Parts Common Post-Processing Methods

(1) Mechanical treatment

In the processing of large print, usually can use a sander or small grinder for simple treatment of the surface of printed parts, and then use more than 800 mesh of sandpaper on its surface for round sanging. Sanding finished, the oil can be sprayed and lubricated using the traditional methed. For most FDM materials, pearlescent treatment is a common post-treatment process. The principle of pearlescent treatment is to achieve the polishing effect through continuous injection of medium beads [3].

(2) Meat treatment

Due to the different properties of different polymer materials, heat treatment processes need to be adopted are also different. At present, there are few experimental data at home and abroad, and appropriate parameters need to be obtained through a large number of experiments [4].

(3) Surface coating treatment

The surface of FDM prints is coated, and the liquid coating is mainly used to fill the gap on the surface to make it flow downstream under the action of surface tension and gravity, so as to improve the surface lubrication [5].

(4) Chemical treatment

After being heated at a high temperature, organic solvent has a good compatibility with its surface. Steam dissolves it on the surface of plastic parts, producing swelling effect and achieving uniform grinding effect. However, most of the solvents used contain toxic substances, so it is not recommended to use for a long time [6, 7].

3. Manual Post-Treatment Process

3.1. Post-Processing Equipment and Tools

Spray pen air pump set with air storage tank, coloring tool set, airtight spray box, coloring pen, solvent bottle, 00000 phase pen, gloves, mask, water diluted solvent, water fill soil spray can, quick dry small fill soil, 80 to 3000 mesh sandpaper, cleaning agent, water paint, non-woven cloth, file, protection paint, water basin, etc.

3.2. The process flow

3.2.1. Model printing. The model was printed and formed by FDM, and the post-treatment model was vase, as shown in figure 2. PLA material was selected, and the printing equipment was Einstart printer produced by hangzhou xianlin 3d technology co., LTD., and the printing parameters were shown in figure 3.

Figure 2. The vase model.

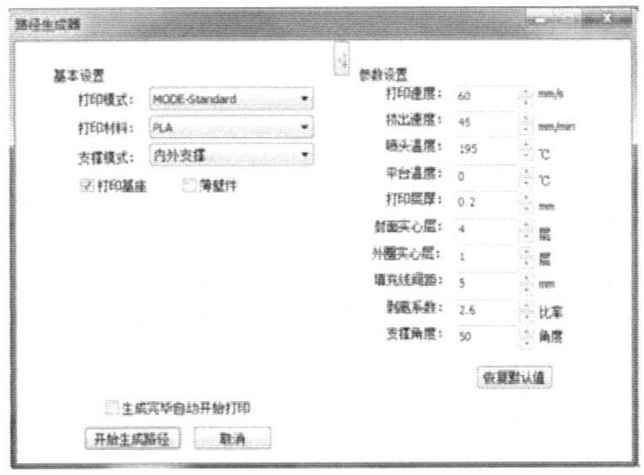

Figure 3. Printing parameters.

3.2.1.1. Remove the support. Once the model is printed, as shown in figure 4, the printing plate is removed and the model is smoothly removed from the bottom plate with a shovel, and then to check whether there is strain concentration model, relatively weak parts with small first stripping knife to spin out the model and the support, and then has a long nose pliers clamping a direction support, applying a constant force, the location of the tiny support can use the file to remove, as shown in figure 5.

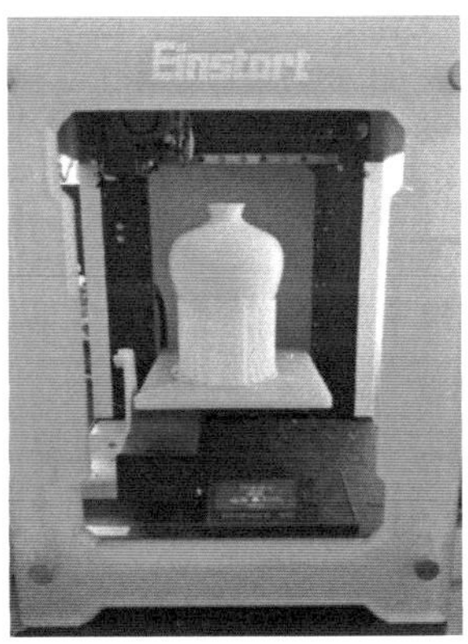

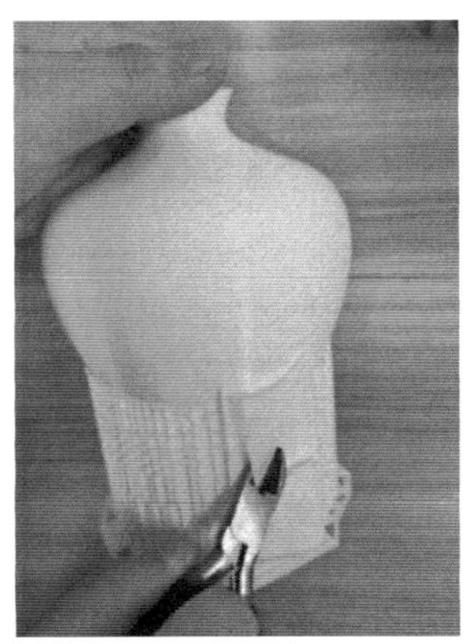

Figure 4. Model finished printing. **Figure 5.** Demolition of support.

3.2.1.2. Low mesh sandpaper polishing. For the "step effect", the model printed by FDM has a rough surface and clear texture. It should be sanded first with a low number of pieces of sandpaper (80-120-240-400-800-1000 mesh). As shown in figure 6, the model and sandpaper are immersed in water and sanded along the model's texture (step texture). Sanding the surface for a period of time if the surface is the same as before, you need to replace a higher number of sandpaper. Wet-water sanding has the advantage of smoothing the surface of the model while extending the lifetime of the sandpaper.

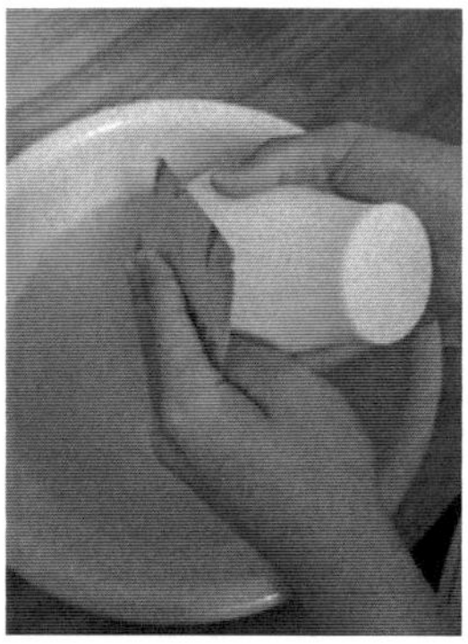

Figure 6. Sand paper for grinding.

3.2.1.3. Quick dry small fill. Apply a small amount of filling material to the gaps or defects in the model, and the fill is daubed quickly and evenly with a hard scraper, as shown in figure 7. Then wait for 30 seconds, after filling soil has hardened, using 1200 mesh to 1500 mesh sandpaper in, as shown in figure 8, If there are still tiny grooves and repeat the above steps. To be in addition to the groove after no large-area fill soil, feel smooth, can proceed to the next step.

Figure 7. Apply small patch of soil evenly.

Figure 8. Polished to make it smooth.

3.2.1.4. Spray can water fill soil spraying. The surface of the model should be washed with water before using the spray pot to fill the soil, wiped with non-woven cloth, dried, and sprayed. The model should be sprayed at the ventilated position. In addition, the model and the nozzle should be kept at about 20cm during spraying, and uniformly and rapidly sprayed 1-3 times, as shown in figure 9. Generally, choose gray spray pot water to fill the soil, because gray is a neutral color, the next step on the color paint less impact.

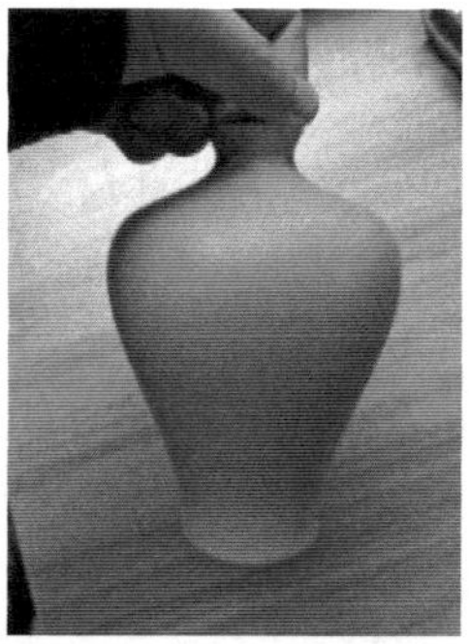

Figure 9. Spray can of water to fill the soil spraying.

3.2.1.5. High mesh sandpaper grinding. After spraying water and filling the soil, place the air outlet. After air drying, use 2000-3000 mesh high-mesh sandpaper for slight grinding. When polishing, pay attention to smooth and uniform grinding along one direction, as shown in figure 10.

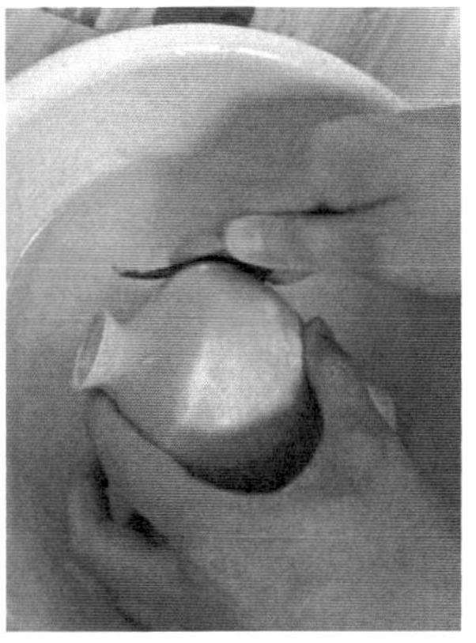

Figure 10. High mesh sand paper grinding.

3.2.1.6. Coloring. Wash and dry the polished and processed models before coloring with pigments. The base color or large area color can be sprayed with a spray gun, as shown in figure 11. The ratio of diluent to pigment should be 1:2 when spraying. The amount of air injection can be adjusted when spraying. Brushes of different thicknesses and sizes can be used to paint the details. It is accessible to use 00000 pens to paint the detailed parts of the figures, or use different widths of the cover tape to cover and then spray the spray gun to paint.

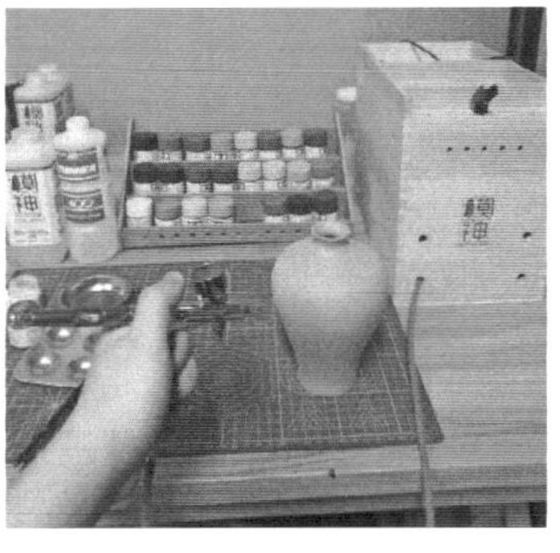

Figure 11. Spray paint.

3.2.1.7. Spray protective paint. After the paint and spray paint are completely dried, protective paint can be sprayed. This model USES B603 water-based extinction, and attention should be paid to uniform spraying, with 2-3 times of external spraying, as shown in figure 12.

3.2.1.8. Complete the work, as shown in figure 13.

Figure 12. Spraying protective pain.

Figure 13. The vase is finished after processing.

Conclusion

First of all, the model should be observed when in printing, reduce unnecessary support as far as possible, when putting the model through software to remove or add support Magics, before printing, printing platform for leveling, a large model, plate coated with a thin layer of printing glue, to prevent buckling deformation, etc. This is the foundation of the late better artificial post-processing.

Secondly, in the manual post-processing should look to the protection work, grinding water mill is the best way to model processing, be patient, 80-2500 mesh, use each mesh sandpaper required time from long to short, low mesh sandpaper grinding along the texture of the model, high mesh sandpaper grinding should be turned around.When mixing colors, you should understand in advance the relationship between light and shade, brightness and purity of various colors, warm and cold color selection, etc.When coming to spray, it ought to be kept at an appropriate distance.

This paper mainly carries out manual post-processing of FDM printing model. According to different models, the degree of difficulty of post-processing method is different, and the process method is also different.

References

[1] Xu R J, Li Z A, Zhu L Y, Li K L and Yang J Q 2016 *Mech. Design Manufact. Eng.* Overview of 3D printing industry and technology development trend, **45(3)** 11-6.

[2] Yan X J, Zhang L and Dong R H, et al. 2018 *Navi. control* The optimization of processing parameters of warp deformation in fused deposition modeling, **17(2)** 85-8.

[3] Zeng F, Yan H S and Wang P 2012 *Mech. Elec. Eng. Technol.* Prototype Production and Post-Processing Technology Based on the FDM, **41(8)** 99-102.

[4] Yan J J 2019 *Shandong indust. technol.* Heat treatment process and technical analysis of metal materials, (16) 22.

[5] Hou G Y, Zhu H, Liu K and Xie D 2017 *Inform. recording mater.* Review of post-processing technology of 3D printed parts, **18(7)** 19-21.

[6] Garg A, Bhattacharya A and Batish A 2016 *Mater. Manufact. Process.* On surface finish and dimensional accuracy of FDM parts after cold vapor treatment, **31** 522-9.

[7] Zhang J T, Hu Z W and Zhu Z P, et al. 2017 *Mech. Sci. Technol. Aerospace Eng.* Experimental studies on spray polishing surface of fused deposition modeling build parts, **36(6)** 960-4. (in Chinese)

Acknowledgments

This research was funded by Young and Middle-aged Scientific Research Backbone Training Program of Nantong Institute of Technology (ZQNGG102，ZQNGG203), The Natural Science Foundation of the Jiangsu Higher Education Institutions of China (18KJA460006, 18KJB460023, 18KJD430006), Priority Discipline Construction Program of Jiangsu Province (2016-9), Qin Lan Project of Jiangsu Province(2016-15,2019-3)and Nantong Science and Technology Commission of China (CP12016002, JC2018144, GY12017022, JCZ18034).

Prismatic Species of Snelson's Tensegrity Structure

Angelo Vumiliya[1], Ani Luo[1], Heping Liu[1] and Andrés González[1]

[1]College of Mechanical and Electrical Engineering, Harbin Engineering University, Harbin, China

vumiliya@hrbeu.edu.cn, luoani@hrbeu.edu.cn, liuheping@hrbeu.edu.cn, andres@hrbeu.edu.cn

Abstract. Snelson's prisms are regular minimal tensegrity structures. They are symmetric according to the rotation axis and have two prisms in two parallel planes. These planes, separated by a certain height, have a finite number of vertices related to the number of bars. Evidently, by changing the position of the upper nodes of bars results in a new stable configuration. This paper demonstrates the possibility to obtain different species through one set and identify the structures' class according to their species.

1. Introduction

The term tensegrity structure refers to a network of tensional and compressive members. The tensional members are designed to be cables, strings or simply elastic material and the compressive are bars or rigid elements. The use of these structures is extensive, as it has appeared in several fields such as aviation and robotics [1-3] because of the controllability, and plays a major role in engineering [4, 5]. However, despite all of these utilities, we only meet one aspect of tensegrity, which is the base of the revolution. Due to its simplicity, a prismatic tensegrity, "Snelson`s prism", has two regular polygons in two parallel planes. They belong to the class $\boldsymbol{k}$ when $\boldsymbol{k}$ bars interconnect; otherwise, they are the first class. It is the base of many structures obtained by connecting several tensegrity units along their axis. These shapes result in domes [6], deployable cylinders [7] by substituting and increasing cables; towers and plates [8] are the result of additional nodes.

Referring to the literature, in [9] a configuration through a single layer and deriving different sets of bars is introduced. The method is general and allows to obtain for a define number of bars an exact stable configuration. Moreover, a mathematical concept was presented in [10] for the same structure within an analytical solution to obtain, for v bars, a v-1 prismatic tensegrity structure. The method refers to planes formed by two bottom nodes and the center of the circle located on top.

In this paper; we show the possibility to obtain different species from an initial set, means j=1; and give out the general formulation of permutation matrix. The method consists in the multiplication and rotation operated on the first configuration.

This paper is organized as follow: firstly, we introduce the Snelson's prims; following, the set of bars where the possibility to obtain different species is labeled. Next, a structure is simulated by variating the twist angle for vertical cables. The conclusion of this work is derived at the end.

2. Snelson's Tensegrity Structure

A Snelson's prism is a set of bars connected through cables. The i-th bar starts from the bottom vertex $\mathbf{n}_i$ and its end is located on the top vertex $\mathbf{n}_{v+i}$. The vertices are located on two parallel planes separated by a defined height h and each plane presents v sides, receiving the name of *prismatic tensegrity*. Hence, the horizontal cables lying on the bottom and upper planes totalize $2v$. Likewise, v vertical cables connect the bottom vertices and the top vertices to stabilize the configuration without any applied load. Therefore, we count $3v$ cables for any Snelson's prism. For example, nine cables are enough to stabilize a 3-bar Snelson's prism "*minimal regular tensegrity prism*" [11].

Table 1. bars and cables connecting vertices of a triangular regular prismatic.

	bars			horizontal cables						Vertical cables		
	$\mathbf{b}_1$	$\mathbf{b}_2$	$\mathbf{b}_3$	$\mathbf{S}_{h1}$	$\mathbf{S}_{h2}$	$\mathbf{S}_{h3}$	$\mathbf{S}_{h4}$	$\mathbf{S}_{h5}$	$\mathbf{S}_{h6}$	$\mathbf{S}_{v1}$	$\mathbf{S}_{v2}$	$\mathbf{S}_{v3}$
bottom vertex	$\mathbf{n}_1$	$\mathbf{n}_2$	$\mathbf{n}_2$	$\mathbf{n}_1$ $\mathbf{n}_2$	$\mathbf{n}_2$ $\mathbf{n}_3$	$\mathbf{n}_1$ $\mathbf{n}_3$		-		$\mathbf{n}_1$	$\mathbf{n}_2$	$\mathbf{n}_3$
upper vertex	$\mathbf{n}_4$	$\mathbf{n}_5$	$\mathbf{n}_6$		-		$\mathbf{n}_4$ $\mathbf{n}_5$	$\mathbf{n}_5$ $\mathbf{n}_6$	$\mathbf{n}_4$ $\mathbf{n}_6$	$\mathbf{n}_6$	$\mathbf{n}_4$	$\mathbf{n}_5$

Therefore, four parameters describe a set of a prismatic tensegrity structure, as seen on figure 1:

- r_b and r_u radii of bottom and top circles, respectively;
- v ($\geq$3) bars which define the polygons of lower and upper circle;
- a height h , which separates the bottom and top circles;
- a twist angle α, relating the projection of the vertical cable lying the top of a bar to the bottom vertex.

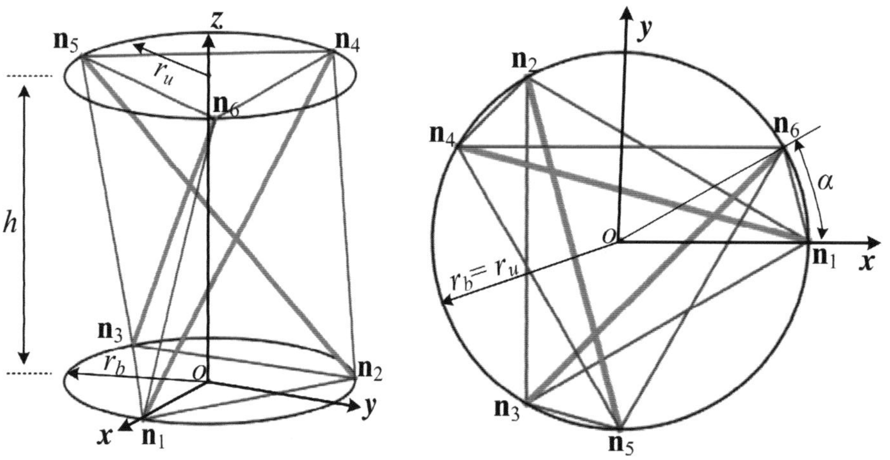

Figure 1. triangular prismatic tensegrity structure, perspective and top view, v=3.

Mathematically, the bottom node and the upper node of the i-th bar are

$$\mathbf{n}_i = r_b\left[\cos 2\pi v^{-1}(i-1) \quad \sin 2\pi v^{-1}(i-1) \quad 0\right]^{\mathrm{T}} \tag{1}$$

$$\mathbf{n}_{v+i} = r_u\left[\cos(2\pi i v^{-1}+\alpha) \quad \sin(2\pi i v^{-1}+\alpha) \quad h r_u{}^{-1}\right]^{\mathrm{T}} \tag{2}$$

in which i=1,2,3, …, v and $\alpha=\pi(0.5\text{-}v^{-1})$. Then, the whole structure geometry matrix is

$$\boldsymbol{N} = \Big[\underbrace{\mathbf{n}_1 \quad \mathbf{n}_2 \quad \cdots \quad \mathbf{n}_i \quad \cdots \quad \mathbf{n}_v}_{\text{bottom vertex}} \;\Big|\; \underbrace{\mathbf{n}_{v+1} \quad \mathbf{n}_{v+2} \quad \cdots \quad \mathbf{n}_{v+i} \quad \cdots \quad \mathbf{n}_{2v}}_{\text{upper vertex}}\Big] \tag{3}$$

From table 1, the bar and cable matrices are

$$\boldsymbol{B}=[\mathbf{b}_1 \quad \mathbf{b}_2 \quad \mathbf{b}_3]=\boldsymbol{N}\boldsymbol{C}_b{}^T; \tag{4}$$

$$\boldsymbol{S}=\Big[\underbrace{\mathbf{s}_{h1} \quad \mathbf{s}_{h2} \quad \mathbf{s}_{h3} \quad \mathbf{s}_{h4} \quad \mathbf{s}_{h5} \quad \mathbf{s}_{h6}}_{\text{horizontal cables}} \;\Big|\; \underbrace{\mathbf{s}_{v1} \quad \mathbf{s}_{v2} \quad \mathbf{s}_{v3}}_{\text{vertical cables}}\Big]=\boldsymbol{N}\boldsymbol{C}_s{}^T \tag{5}$$

where $\boldsymbol{C}_b{}^T$ and $\boldsymbol{C}_s{}^T$ are bar and cable connectivity matrices given as

$$\boldsymbol{C}_b{}^T=\begin{bmatrix}-\boldsymbol{I}_v\\ \boldsymbol{I}_v\end{bmatrix};\ \boldsymbol{C}_s{}^T=\begin{bmatrix}\underbrace{\begin{matrix}\boldsymbol{E}_1 & \boldsymbol{O}\\ \boldsymbol{O} & \boldsymbol{E}_1\end{matrix}}_{C_h{}^T} & \underbrace{\begin{matrix}-\boldsymbol{I}_v\\ \boldsymbol{E}_2\end{matrix}}_{C_v{}^T}\end{bmatrix} \tag{6}$$

$\boldsymbol{I}_v$ and $\boldsymbol{O}$ are $v \times v$ identity and zero matrix. $\boldsymbol{E}_1$ and $\boldsymbol{E}_2$ are

$$\boldsymbol{E}_1=\begin{bmatrix}-1 & 0 & 0 & 0 & 1\\ 1 & -1 & 0 & \cdots & 0\\ 0 & 1 & \ddots & \cdots & 0\\ 0 & 0 & \ddots & -1 & 0\\ 0 & 0 & 0 & 1 & -1\end{bmatrix}_{v\times v} \text{ and } \boldsymbol{E}_2=\begin{bmatrix}\boldsymbol{O}_{v-1\times 1} & \boldsymbol{I}_{v-1}\\ 1 & \boldsymbol{O}_{1\times v-1}\end{bmatrix}_{v\times v} \tag{7}$$

respectively. For any prismatic tensegrity structure, the sets of geometry and typology parameters remain the same.

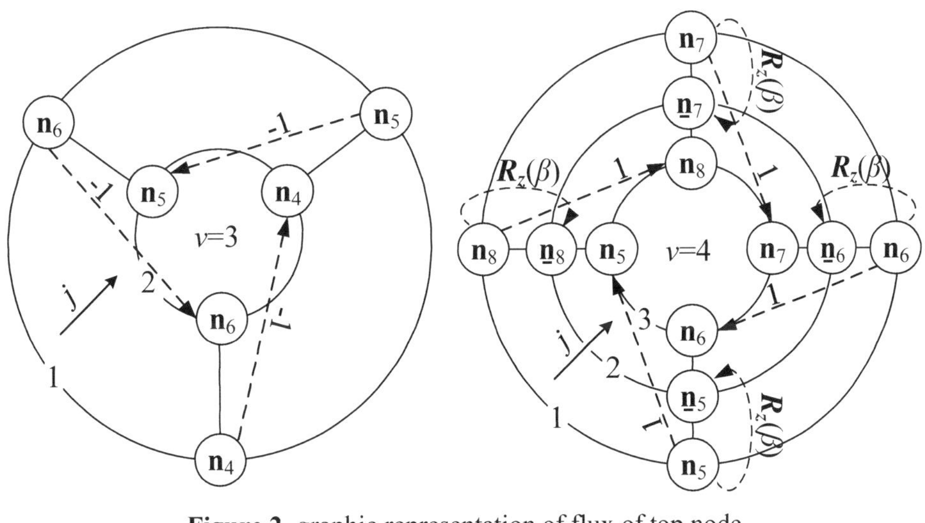

Figure 2. graphic representation of flux of top node.

3. Species of a Prismatic Tensegrity Structure

They are expressed in terms of the top nodes [10]. One bar can be connected to any bottom node except its corresponding node. Therefore, for v bars exist more than one structure. Due to the permutation of top nodes, we can reach another structure using multiplication operations. The bottom nodes are duplicated while the top nodes keep changing the position as seen on figure 2. Let's consider matrix $\boldsymbol{P}_1$

and $\boldsymbol{P}_2$ are divided into two parts. The first one is the duplicate of the bottom nodal coordinates and the second one is related to j different species.

$$\boldsymbol{N}_j = \begin{cases} \boldsymbol{N} & \text{if } j=1 \\ \boldsymbol{N}_{j-1}\boldsymbol{P}_1 & \text{if } j\neq 1 \text{ and } v \text{ odd} \\ \text{for } v \text{ even} \begin{cases} \left[\boldsymbol{N}_{[3\times v]} \middle| \boldsymbol{R}_z(\beta)\boldsymbol{N}_{[3\times v+1,\cdots,2v]}\right] & \text{if } j=2 \\ \boldsymbol{N}_{j-2}\boldsymbol{P}_2 & \text{otherwise} \end{cases} \end{cases} \tag{8}$$

in which

$$[\boldsymbol{P}_1]_{2v\times 2v} = \left[\underbrace{\begin{matrix} \boldsymbol{I}_v \\ \boldsymbol{O}_{v\times v} \end{matrix}}_{2v\times v} \middle| \underbrace{\begin{matrix} \multicolumn{2}{c}{\boldsymbol{O}_{v\times v}} \\ \boldsymbol{O}_{v-\frac{v-1}{2}\times\frac{v-1}{2}} & -\boldsymbol{I}_{v-\frac{v-1}{2}} \\ -\boldsymbol{I}_{\frac{v-1}{2}} & \boldsymbol{O}_{\frac{v-1}{2}\times v-\frac{v-1}{2}} \end{matrix}}_{2v\times v} \right]; \quad [\boldsymbol{P}_2]_{2v\times 2v} = \left[\underbrace{\begin{matrix} \boldsymbol{I}_v \\ \boldsymbol{O}_{v\times v} \end{matrix}}_{2v\times v} \middle| \underbrace{\begin{matrix} \multicolumn{2}{c}{\boldsymbol{O}_{v\times v}} \\ 0 \ \cdots \ 0 & 1 \\ & 0 \\ \boldsymbol{I}_{v-1} & \vdots \\ & 0 \end{matrix}}_{2v\times v} \right] \tag{9}$$

The matrix $\boldsymbol{R}_z(\beta)$ gives a set of top nodes rotated about z-axis an angle $\beta=\pi v^{-1}$.

4. Connectivity Matrix of Species

The related bar and cable matrix are

$$\boldsymbol{B}_j = \boldsymbol{N}_j\boldsymbol{C}_b^{\boldsymbol{T}}; \tag{10}$$

$$\boldsymbol{S}_j = \boldsymbol{N}_j\boldsymbol{C}_{sj}^{\boldsymbol{T}}; \tag{11}$$

where the connectivity matrix $\boldsymbol{C}_{sj}^{T}$ has the following form

$$\boldsymbol{C}_{sj}^{\boldsymbol{T}} = \begin{bmatrix} \underbrace{\begin{matrix} \boldsymbol{E}_1 & 0 \\ 0 & \boldsymbol{E}_1 \end{matrix}}_{C_h^{T}} & \underbrace{\begin{matrix} -\boldsymbol{I}_v \\ \boldsymbol{E}_{2j} \end{matrix}}_{C_{vj}^{T}} \end{bmatrix} \tag{12}$$

Then;

$$\boldsymbol{E}_{2j} = \begin{cases} \boldsymbol{E}_2 & \text{if } j = 1 \\ \boldsymbol{E}_{2j-1}\boldsymbol{E}_2 & \text{otherwise} \end{cases} \tag{13}$$

5. Simulated Results and Conclusion

The results obtained are in agreement with other literature [9, 10]. The success is the obtention of the twist angle of a vertical cable related to self-equilibrium. It defines also the class of tensegrity. The tensegrities with odd number of bars are class one. Differently to the structure constructed with even number of bars, when $j=v/2$, the structure is a v-th class tensegrity structure.

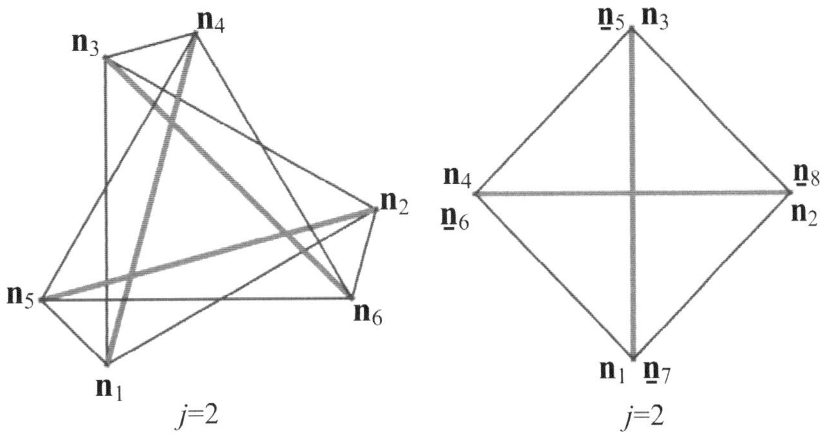

a) Top view of a class 1, v=3　　b) Top view of a class 4, v=4

Figure 3. simulated prismatic tensegrity structure.

Table 2. twist angle of vertical cables.

v \ α_j \ j	1	2	3	4	5	6
3	30	-30	-	-	-	-
4	45	0	-45	-	-	-
5	54	18	-18	-54	-	-
6	60	30	0	-30	-60	-
7	64.2857	38.5714	12.8571	-12.8571	-38.5714	-64.2857

The method for constructing different species of prismatic tensegrity structures has been achieved. Different matrix, geometry matrix and typology matrix, are simply obtained from the first configuration. In addition, the bar and vertical cables connectivity matrix remains the same for any configuration while the vertical cables connectivity matrix are multiplied by itself for next configuration.

References

[1] Sultan C and Skelton R 2003 *Int. J. Solids Struct.* Deployment of tensegrity structures, **40(18)** 4637-57.
[2] Luo A, Wang J and Liu H 2017 *2017 IEEE Int. Conf. Mechatronics Autom. ICMA 2017* Four-bar tensegrity robot based on ADAMS simulation, 1463-68.
[3] Luo A and Liu H 2017 *J. Mech. Robot.* Analysis for Feasibility of the Method for Bars Driving the Ball Tensegrity Robot, **9(5)** 051010.
[4] Veuve N, Dalil Safaei S and Smith I F C 2016 *Eng. Struct.* Active control for mid-span connection of a deployable tensegrity footbridge, **112** 245-55.
[5] Averseng J and Dubé J F 2012 *Procedia Eng.* Design, analysis and self stress setting of a lightweight deployable tensegrity modular structure, **40** 14-9.
[6] González A, Luo A and Liu H 2017 *IASS 2017* Study of a novel design for self-standing tensegrity domes, **1** 1-8.
[7] Liu H, Geng J and Luo A 2017 *Compos. Struct.* Tensegrity configuration method for connecting tensegrity units along their axes, **162** 341-50.
[8] Nagase K, Yamashita T and Kawabata N 2016 *Mech. Res. Commun.* On a connectivity matrix formula for tensegrity prism plates, **77** 29-43.

[9] Luo A, Liu H, Skelton R E and Che S 2017 *Chinese J. Mech. Eng.* The theory of basic tensegrity unit stable forming, **53(23)** 62-73.
[10] Vumiliya A, Luo A and Song Z 2018 *2018 IEEE Int. Conf. Mechatron. Autom. (ICMA)* Mathematic Modeling for Configuration of Tensegrity Structure, 104-9.
[11] De Oliveira M C and Skelton R E 2009 Tensegrity systems, Springer.

Acknowledgement
This project is supported by National Natural Science Foundation of China (Grant No. 51605111, 51675114, 51875111).

IOP Conf. Series: Materials Science and Engineering **649** (2019) 012014 doi:10.1088/1757-899X/649/1/012014

Air to Air Heat Exchanger Mimicked from the Directional Passive Water Transport Featured by the Skin Morphology of Phrynosoma Cornutum

SÁNCHEZ Federico[1] and ZHU Shifan[1]

[1]College of Mechanical and Electrical Engineering, Harbin Engineering University, China

federico.sanchez@ucr.ac.cr, zhushifan@hrbeu.edu.cn

Abstract. A biomimetic approach is proposed to improve the performance of an intercooler using only the passive water transport featured by the lizard species Phrynosoma Cornutum for racing applications. The lizard is able to transport in a preferred direction, water over its integument whilst consuming no energy to do so. That feature is used in the external surface of the intercooler and an evaporation rate is desired. The model started with the air conditions after the turbo charger stage, Then, the convection of air in the heat exchanger is modelled by the boundary layer theory which is applicable for the scale used justified by the Péclet number value, for both the heat and mass transfer. The convection heat transfer was satisfactory modelled and the values reached inside the expected values for the optimal 2 mm wall thickness selected to demonstrate the enhancement due to the mass transfer and a surface temperature of 62.02 °C was stabilized for the conditions selected under a few iterations. According to the mass convection theory a mathematical model of the heat exchange was made and for the heat lost there exist a drawback due to a deviation from real values due to the highly turbulent flow under the conditions selected and the very small value of the Schmidt number obtained in the literature and pooled to the lack of a proper equation to compute the Sherwood number, which was not found and modelled outside the Schmidt number range. To solve this, a correction factor of the water evaporation is introduced which assumes that not all the water from the channels evaporates and some part is taken by the turbulent air as water droplets, but it needs further validation by real experiments, which is expected to be the next stage of the project.

1. Introduction

Natural systems have been existing over 3.8 billion years [1] of evolution, natural selection and constant improvement, when compared with human knowledge, the latter becomes insignificant. Life forms a technology in every proper sense, with a diversity of design, materials, engines, and mechanical contrivances of every degree of complexity [2]. Back in 1997, a woman spread and popularized the term Biomimicry [3] when she wrote her book and stablished a completely new field of research, by which many professionals including a wide range of experts were able to solve their problems while working interdisciplinary. The main idea is bring not only biologists but nature's designs to the design table, while seeking sustainability and innovation [4].

According to Gleich et al. [5] Biomimetics have been developed in three main strands: the first and oldest one is functional morphology, form and function by which it's favored the structure rather the material properties; the second strand is about signal and information processing, bio cybernetics, bio inspired sensors and robots; the third and newer one is nanotechnology by nanomimetics and molecular self-organization.

Biomimetics is the attempt to learn from nature; it deals with the development of innovations on the basis of investigation of natural, evolutionary optimized biological structures, functions processes and systems [5].

Research and development in biomimicry and biomimetism is measured by the Da Vinci Index 2.0, created in 2013 by the Fermanian Business & Economic Institute located in Point Loma Nazarene University in San Diego, California [6]. It is an improvement from the first Da Vinci Index created two years before by the same institution, based in number of scholarly articles, patents, grants and the dollar value of those grants. Innovation and sustainability and the main drives of these research areas [1]. In short, *biomimicry can be defined as the innovation inspired by nature.* [3, 4, 7] Some websites offer online solutions such as Biommicry Institute [4], AskNature.org and the Sustainability Workshop of AutoDesk® [7] based on past user experience or designs.

Since the invention of the turbo engine systems, its functions have been always the same: to inject more fresh and compressed air into the engine in the same time lapse and to improve the peak performance of both power and torque output.

While the air is compressed its density and temperature raise accordingly to levels not really safe for the engine block because it will increase the temperature of the components of the engine, thus the cooling system will have to work harder in order to maintain a safe operation and to obtain the output values for which it was designed for.

Heat exchangers, called intercoolers have been used to chill the air temperature before entering the intake manifold preventing the phenomenon of engine knocking or pre-detonation of the fuel-air mixture inside the combustion chamber. This means that the fuel will ignite instants before the spark plug detonates it at the precise moment it was intended for. This precise moment is determined by the setup and tuning of the ECU (Engine Control Unit) which is a programmable chip that controls the timings and values of the engine. This knocking will damage the engine almost immediately because it counteracts the angular momentum of the crankshaft by pushing the piston against its angular velocity trying to stop it suddenly, acting as an engine brake stroke.

Intercoolers are radiators (more generally heat exchangers) that cool down the air that is going straight to the intake manifold of the engine the same way the radiator of the cooling system maintains the engine within its proper, normal and safe operating range of temperatures. The most common types of intercoolers are, air to air, air to water and the merge of those to types for this kind of application. It should be emphasized that what is cooled down inside the intercooler at any moment is the intake air before being injected to the engine, opposite as the radiator, where what is cooled is the coolant circulating the engine to maintain certain operating range of temperature.

To improve the effectiveness of cooling the air temperature within the intercooler using the same device model and the same time for the process, some car manufacturers like Subaru® [8] take advantage of a fine pulverized cold water stream, sprayed all over the front outer surface of the intercooler where a fresh stream of laminar flow air is passing through the space between the intercooler ducts at high speed. This water will extract some of the energy in form of heat from the hot and compressed air, raising its temperature until it reaches the saturation temperature of water at that specific pressure (depending on the atmospheric conditions where the intercooler is located) to start the evaporation process as a saturated mixture of water and water vapor.

Intercoolers work by means of forced external convection heat transfer occurring at the outer face of the heat exchanger. The stream of air is renewed constantly due to the movement of the car, taking away heat faster. The evaporation of water will take even more heat which will be sent far away from the intercooler by the same fast air stream, producing forced convection.

The injection of the pulverized water spray needs a high pressure pump, pipes and nozzles to achieve the proper pulverization and not just to leak water (which will be useless). This water conditions require to extract some power from the engine to drive the pump into the designed pressure of the system, causing a tiny loss of power in the engine from that extra power generated thanks to the system. At the end, the overall gain of power is more than that used to move the water injection system.

2. Methodology

2.1. Lizard Skin

The water over the Texas Horned Lizard (Phrynosoma Cornutum) flows at almost doubled velocity to the snout than to the tail in comparison. It was reported by Comanns et. al. in [9] that the within the first 333 ms, after the water was in contact with the skin, the velocity towards the snout was $3.15 \pm 0.94\ mm/s$ while the velocity towards the tail was $1.61 \pm 0.45\ mm/s$.. The passive water transport is possible over the lizard's integument because of the morphology of the integument itself or skin, where there are periodic and asymmetric changes in the shape leading to abrupt widening's and gradual narrowing's of the capillary canals, this is one on the essential features of the skin. The second essential feature is: there are some interconnections between the channels in order to continue the halted stream of water when it reaches an abrupt widening [10], as can be clearly seen in Figure 3.

In the real integument of the Phrynosoma Cornutum, the dimensions of the capillary channels ranges from 30 μm to 300 μm. This phenomenon is possible due to the super hydrophilic surface the scales of the lizard has. It was said in the last section that the practical angle to separate hydrophobicity from hydrophilicity is 65°. The angle of contact between the water and the scales is smaller than 10° which makes the skin of the lizard super hydrophilic [10]. This feature improves the movement of water because it doesn't let the skin to get wet while minimizing the loss of water due to evaporation.

2.2. Fluid as Ideal Gases

The saturated water vapor expected to be produced and the air, should be modelled, and the ideal gas is preferred over the real gases for obvious reasons. Nevertheless, one of the simplest methods available in the literature is the *compressibility factor Z* that is a measure of the deviation from the ideal gas behavior suffered by all real gases. Real gases deviate severely from the ideal behavior near the critical point and also close to the saturation state in the specific volume vs. temperature diagram or thermodynamic v-T phase diagram. For the case of real gases, "*gases behave differently at a given temperature and pressure, but they behave very much the same at temperatures and pressures normalized with respect to their critical temperatures and pressures. The normalization is done as*" [11]. At low pressures regardless of the temperature of the gas, it behaves as if it were an ideal gas. At high temperatures, the ideal gas can be assumed to be accurate enough to be used as a model. Obviously the closer the state of a gas is to its critical point, greater will be the deviation of the ideal gas behavior due to the liquefaction of the gas at high pressure.

A simple explanation of the temperature of the critical point can be given by the liquefaction of gases, which accounts for the most of the deviation of real gases from the ideal gas behavior and "*if a gas is below its critical temperature, it can be liquefied simply by compression. However, the critical temperature for most gases is very low*" [12].

2.3. Turbo Compressor Stage

Now a way to obtain the temperature of the air after the compressor stage of the turbo should be implemented and the mathematical model used for the air of the air after the turbo compressor stage are given by [13]. Here is where we make use of the specific characteristic curve of an actual turbo given by a manufacturer [14], because data is required to input in the equation of temperature, pressure and mass flow of air after being compressed.

$$T_{out,turbo} = T_{in} + \frac{T_{in} \cdot [-1 + (P_{out}/P_{in})^{0.263}]}{Efficiency} \tag{1}$$

2.4. Heat Transfer for the Case of a Dry External Intercooler Surface

It is well known that intercoolers are air to air heat exchangers that work by means of internal and external convection under cross flow with no stream mixture at any time. This makes it very difficult to model accurately without the use of the empirical models given by [16, 17], and [15]. Let us remember the Nusselt number that compares the heat transfer enhancement of convection with respect to the conduction phenomenon for the same temperature conditions of the wall and the fluid. L_c is the characteristic length of the problem.

Once the Nusselt number is obtained and the variables of state of the air, both inside and outside the intercooler duct are known, the temperature values of the air at the outlets of the external and internal air streams are computed by means of the Logarithmic Mean Temperature, and the equivalent thermal resistance method, using the modified Newton law of convection.

$$\dot{Q} = UA_s\Delta T_{lm} \tag{2}$$

Where $\Delta T_{lm}$the usual logarithmic mean temperature difference and U is the heat transfer coefficient of convection obtained by equivalent thermal resistance method.

$$\Delta T_{lm} = \frac{\Delta T_1 - \Delta T_2}{\ln(\Delta T_1/\Delta T_2)} \tag{3}$$

The correction factor F for cross flow heat exchangers (inlet and outlet values of temperature of both hot and cold streams. It basically computes a transformation of the equivalent temperature difference of the cross flow to compare with the parallel flow, like a scale or conversion factor.

$$\Delta T_{lm} = F\Delta T_{lm,CF} \tag{4}$$

The limiting value for F is 1, this means $F \leq 1$. So it is a normalization of the deviation from the counter flow case, similar to the ideal and real gas models difference. Two temperature ratios must be calculated to use *F*, *P* and *R*:

$$R = \frac{T_1 - T_2}{t_2 - t_1} = \frac{(\dot{m}c_p)_{tube\ side}}{(\dot{m}c_p)_{shell\ side}} \tag{5}$$

$$P = \frac{t_2 - t_1}{T_1 - t_1} \tag{6}$$

Where the subscript 1 represents the inlet pipe and the 2 the outlet for both hot and cold air streams, T represents the shell side (or the outside air stream) and t represents the tube (or inside air stream) temperatures in equation (5) and equation (6). Relating those expressions the correction factor F can be found using any correlation factor graphic obtained easily in [16, 17], and [15].

2.5. External Forced Convection

For the case of a square of edge length D duct with forced flow under external convection and when the work fluid is a gas, in our case air, the Nusselt number is:

$$Nu = 0.094Re^{0.675}Pr^{1/3} \tag{7}$$

For equation (7), its validity is conditioned of the case when $3{,}900 < Re < 79{,}000$. Re is the Reynolds number, which is a ratio of the inertial forces of the fluid over the viscous forces and Pr is the Prandtl number, which is a ratio of the dependence of the molecular diffusivity of momentum or kinematic viscosity ($\nu=\mu/\rho$) over the molecular diffusivity of heat ($\alpha=k/\rho c_p$). The diameter used for is the hydraulic diameter for a rectangular shape.

2.6. Internal Forced Convection

To obtain the Nusselt number for the case of *turbulent flow* under internal forced convection is a modification of the second Petukhov equation improved for better accurate calculations and also is able to model the behavior for lower Re values. The range of application of the modified Petukhov equation is $3{,}000 < Re < 5{,}000{,}000$ and $0.5 \leq Pr \leq 2{,}000$.

$$Nu = \frac{(f/8)(Re - 1000)Pr}{1 + 12.7\,(f/8)^{0.5}\,(Pr^{2/3} - 1)} \tag{8}$$

This are all that is needed to obtain all the possible cases of the convection of air concerning the air to air intercooler. It should be emphasized that so far we have only calculated the convection of air, but we also need to use the water evaporation over the external surface of the heat exchanger in the next section. Again in equation (8), Re is the Reynolds number, Pr is the Prandtl number and f is the friction factor obtained by the usual Blasius equation available in any piping textbook.

2.7. Evaporative Heat Transfer and Energy taken from the Water Mass Transfer

Once the model of the dry surface intercooler is set up with the well known *boundary layer theory*, let us consider the case at which the channel diameter transporting compressed water over a "flat surface" may be small enough in dimension to not let the macroscopic boundary layer theory to be valid due to the size, because the boundary layer will be of the same order of magnitude as the hydraulic diameter of the channel.

As the channel get smaller in dimension, the mass diffusion forces will gain importance and may overcome the boundary layer forces. This means that a ratio of the momentum (and convection) diffusivity of the mass transfer and the diffusion of the mass must be quantified using the Péclet number [16]

$$Pe = \frac{vL_c}{D_{water-air}} \tag{9}$$

where v is the velocity of flow, L_c the characteristic length and $D_{water-air}$ is the diffusion coefficient called Marrero and Mason formula, widely spread over the literature.

The use of Pe is justified and "*based on the mean flow velocity through the narrowest opening between the tubes*" [15], or in this case the narrowest part of the open channel carved over the flat surface of the heat exchanger, because of the periodic changes in the geometry allowing the passive water transport in the preferred direction.

For the dimensions of the channel with small values of length and of course Pe, there exists a dominant region where the effects are felt, and "*In this region of small Péclet-numbers where diffusion and convection are of equal importance, few correlations are available for flat structures under laminar flow*" [15].

In order to solve this, two cases of the Sherwood number are solved numerically and presented by Kachel et. al. resulting in one expression for the case of a macroscopic channel carved in a flat plate that is big enough, this means that is the region where the **boundary layer is applicable**, Sh is for the case of laminar flow over a flat surface

$$Sh = 0.664\left(\frac{v}{D_{water-air}}\right)^{-1/6} Pe^{1/2} \tag{10}$$

where v is the dynamic viscosity and D is the diffusion coefficient of water in air.

In the contrary case of turbulent flow over the flat surface and when the boundary layer theory is still valid, equation

(10) must not be used and instead the equation

(11) is needed to evaluate the value of Sh, for Sc > 0.6

$$Sh = 0.037Re_L^{0.8}Sc^{1/3} \tag{11}$$

IOP Conf. Series: Materials Science and Engineering **649** (2019) 012014 doi:10.1088/1757-899X/649/1/012014

Where Sc is the Schmidt number, the ratio of the kinematic viscosity and the diffusion of water vapor in air, from equation

$$Sc = \frac{\nu}{D_{water-air}} \tag{12}$$

If the dimension of the channel is small enough in the order of magnitude of 10^{-5} m, that is the case of Kachel et. al. ***Pe* → 0 and then the macroscopic boundary layer theory is no longer applicable**, because it neglects the fact that the diffusion forces are always present, but those forces a diminished by the fact of the convective forces are much stronger. then the Sherwood number becomes numerically [16], for the case of laminar flow:

$$Sh = 0.8\, Pe^{0.1} \tag{13}$$

For the case of small Péclet number under turbulent conditions there are no expression available in the literature.

To find the coefficient of convection due to the mass transfer, h_{heat} the mass convection h_{mass} should be known first. From the definition of the Sherwood number which a measure of the convection heat mass transfer and the diffusive mass transfer.

$$h_{heat} = \rho c_p h_{mass} \left(\frac{\alpha}{D_{water-air}}\right)^{2/3} \tag{14}$$

With this expression, and introducing it into equation (2) the model is completed for the case of the dry intercooler surface when the water transport and the evaporation are occurring simultaneously.

3. Final Design

The material selected of the final design is Al 6063-O, a heat treated alloy with good thermal conductivity and proper elasticity modulus suitable for this application. The final dimensions of the intercooler are 600 mm by 60 mm or all ducts and a height 15 mm with a separation of 15 mm with a total of 6 ducts. The thickness selected is 2 mm for all the walls. For the selected pressure boost of 2 bar absolute, the safety factor is 1.8, as can be seen in Figure 1 and Figure 2.

Figure 1. Proposed design of the intercooler with the specified dimensions.

Figure 2. Von Mises Stress pattern for the case of 2 mm thickness. Remember E = 71.7 [GPa].

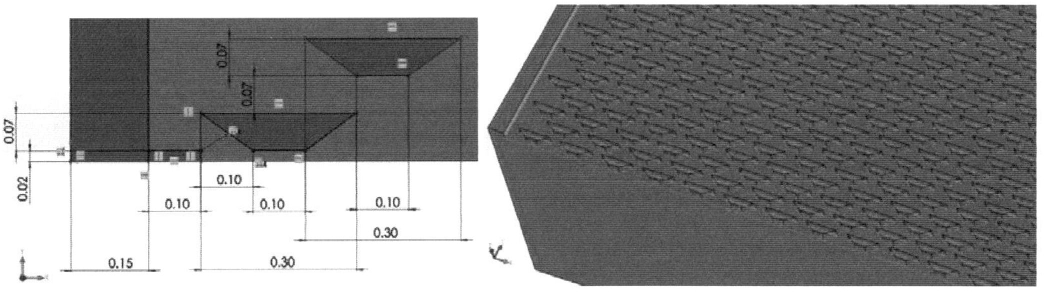

Figure 3. Mimicked morphology from the Texas Horned Lizard. Dimessions in [mm] and actual look of the improved model (not calculated in this paper) but the morphology is periodically as well.

Results

With the model of the intercooler in mind, an external temperature of air of 20 °C and a pressure of 1 bar were selected, with a HR of 70%. For all the cases used, the air and the water vapor were treated as ideal gases because of the criteria stated above as the Z factor in all cases was very close to the unity. A volumetric flow of 0.15m³/s was selected to be compressed by the turbo, and from equation (1) a discharge temperature of 109 °C was found, which is the temperature of the inlet of the intercooler. Then the mass flow rate is 0.283 kg/s and the volumetric flow is 0.0279 m³/s. Under these conditions, the viscosity and diffusivity are 1.93×10^{-5} m²/s and 2.27×10^{-5} m²/s respectively. For the internal convection then, from the dimensions given above and the air state, the velocity inside the duct is 7.55 m/s, for which corresponds a Re of 7215 which was practically treated as the turbulent flow was fully developed near the entrance. The Pr is 0.71 and the friction factor is found to be 0.034.

With all those numbers, the Nu is 22. 728 per tube and we have six tubes. Finally using the Petukhov equation, the value of h_{in} was found to be 230 W/m²·K.

For the case of external convection, the velocity of air was assumed to be 25 m/s. Then, the viscosity and diffusion are 1.52×10^{-5} m²/s and 2.07×10^{-5} m²/s. Therefore, Re is 39577 (turbulent) and Pr is 0.731. Nu was calculated with equation (7) and it was 107. Finally, h_{out} is 114 W/m²·K.

After a few iterations, the model converged to an external surface temperature of the intercooler of 62 °C and a discharge temperature of 65 °C, dissipating 2625 W, for dry conditions.

Table 1. Temperatures for both air streams obtained after several iteration of the model ran into an Excel ® spreadsheet with the model of the heat exchanger.

Temperature (°C)	T_{in}	T_{out}
Hot air inside intercooler	109	65
Cold air outside intercooler	20	24.2

For the case when water is being transported passively over the surface, Re is now 48 265, Le is 0.649, Sh is 50 and Pe is 1565 which means that the boundary layer theory is applicable. From the Marrero and Mason equation, the water diffusion on air is 3.193×10^{-5} m²/s.

Sc for those conditions is 0.0138, which is well below the range of applicability of equation (11). For that, an efficiency factor of evaporation was introduced which model how much of the water over the surface evaporates, and the rest is taken by the turbulent air in form of droplets, which requires further validation and simulation in the next stage of the project. An efficiency of 50% was used, and the lead to a mass convection of 0.054 m/s calculated from the definition of Sh. With this, the mass flow is 0.0033 kg/s of water over the surface and the convection coefficient of heat is 49.2 W/m²·K. This makes the surface temperature converge at 45.3 °C, dissipating almost 150 W extra. This obviously is well below the expected temperature drop to be around 10 °C, so results are taken with care and by no means are lead to a final decision.

Table 2. Final results from the calculation of the heat transfer and water evaporation model with the iteration stopped at $T_{h,out} = 65$ °C.

Case	$\dot{Q}$ [kW]	T_s [°C]
Convection only	2.62	62.05
Water evaporation and convection	0.15	45.31

Conclusions

According to the mass convection theory, for the heat lost there exist a drawback due to a deviation from real values due to the highly turbulent flow under the conditions selected and the very small value of the Schmidt number obtained in the literature and pooled to the lack of a proper expression to compute the Sherwood number under macroscopic turbulent flow, which was not found and it was modelled outside the Schmidt number range. This means that the model of water evaporation and mass transfer should be improved.

However, the model proposed was satisfactory, it needs further validation and extensive simulations due to the lack of literature concerning the evaporation of water over the lizard skin. Also this is only the model and the first stage of subsequent works.

The water evaporation allowed to increase the convection heat exchange by adding a convection of mass to the convection of heat. However, from results and real data the values obtained were inside the range expected.

Also, due to dynamic condition in the functioning of an engine, real data are to be collected from experiments in the next stage of the project.

Summary

Péclet and Schmidt numbers are given, which separate the boundary layer theory and the diffusion theory and stablishes clear limits for both cases and when they overlap. Basically the boundary layer theory is not valid if Péclet number is smaller than unity. Then the Sherwood number is introduced to find the mass convection for turbulent and laminar flows, and is analogous to the Nusselt number for mass convection and diffusion.

The aluminum is chosen because of the mechanical properties such as weight and weight and the thermal properties, such as the conduction makes it ideal for this application. It is justified also why there is no calculation of the stress, because the projects focused on the heat transfer and not in the mechanical design, but a simulation was made with the optional wall thicknesses just to be sure everything is under safe functioning.

Then is found the Schmidt and Sherwood numbers and it is found that for our conditions of very small Schmidt number and highly turbulent flow, there exist no expression in the literature for the Sherwood number to find the exact mass convection, even though many scientific papers were consulted on the topic and not used in the references. Then a correction factor is used because from experience there can't exist a very big value of temperature decrease and the evaporation efficiency is introduced as a normalized percentage of the water that actually evaporates and the rest is taken away from the channels by the fast and turbulent air.

Reference

[1] Helfman-Cohen Y and Reich Y 2016 Biomimetic Design Method for Innovation and Sustainability, Springer.
[2] Vogel S 1998 Cat's Paws and Catapults: Mechanical Worlds of Nature and People, New York: W. W. Norton & Company.
[3] Benyus J 1997 Biomimicry: Innovation Inspired by Nature, New York: Quill.
[4] B. Institute, "AskNature," Biomimicry Institute, 2018. [Online]. Available: https://biomimicry.org. [Accessed 7 5 2018].
[5] Gleich A V, Pade C, Petschow U and Pissarskoi E 2010 Potentials and Trends in Biomimetics, Heidelberg: Springer.
[6] F. B. &. E. Institute, "pointloma.edu," Point Loma Nazarene University, 2011. [Online]. Available: https://www.pointloma.edu/centers-institutes/fermanian-business-economic-institute/da-vinci-index-biomimicry. [Accessed 4 5 2018].
[7] Tavsan F and Sonmez E 2015 *Procedia - Social Behav. Sci.* Biomimicry in Furniture Design, **197** 2285-92.
[8] AutoDesk, "Sustainability Workshop Autodesk," AutoDesk, [Online]. Available: https://sustainabilityworkshop.autodesk.com/products/doing-biomimicry-mechanical-principles. [Accessed 9 5 2018].
[9] Subaru, "Subaru Tecnica Internacional," Subaru Starlink, 2018. [Online]. Available: https://www.subaru.com/2019-s209. [Accessed 15 01 2019].
[10] Comanns P, Buchberger G, Buchsbaum A, Baumgartner R, Kloger A, Bauer S and Baumgartner W 2015 *The Royal Soc. Interface* Directional, passive liquid transport: the Texas horned lizard as a model for a biomimetic 'liquid diode' **12(109)**.
[11] Comanns P, Effertz C, Hischen F, Staudt K, Böhme W and Baumgartner W 2011 *Beilstein J. Nanotechnol.* Moisture harvesting and water transport through specialized micro-structures on the integument of lizards, **25** 204-2014.
[12] Cengel Y A and Ghajar A J 2015 Thermodynamics an Engineering Approach, New York: McGraw-Hill Education.
[13] Matveev A N 1985 Molecular Physics, Moscow: Mir Publishers.
[14] N. A. S. I. O. Club, 2010 A Complete Tuning Guide, North American Subaru Impreza Owners Club.
[15] G. M. INC., "www.garrettmotion.com," GARRETT MOTION INC., 2019. [Online]. Available: https://www.garrettmotion.com/racing-and-performance/performance-catalog/turbo/gt3076r/. [Accessed 27 02 2019].
[16] Cengel Y A and Ghajar A J 2015 Heat and Mass Transfer, New York: McGraw Hill Education.
[17] Webster J G and Eren H 2014 Heat flux, in Measurement, instrumentation and sensors handbook. Spatial, mechanical, thermal and radiation measurement, Boca Raton, CRC Press, **67** 1-15.
[18] H. L. IV J and H. L. V J 2008 A heat transfer textbook, Cambridge, MA: Phlogiston Press.
[19] Kachel S, Zhou Y, Scharfer P, Vrančić C, Petrich W and Schabel W, 2014 *The Royal Soc. Chem.* Evaporation from open microchannel grooves, **14(14)** 771-8.

Acknowledgements

This paper is supported by Heilongjiang Science Foundation, Grant No.QC201603 and the International Exchange Program of Harbin Engineering University for Innovation-Oriented Talents Cultivation.

IOP Conf. Series: Materials Science and Engineering **649** (2019) 012015 doi:10.1088/1757-899X/649/1/012015

Multi-stable Deployment of Unitary Tensegrity Structures

Andrés González[1], Ani Luo[1] and Dongyan Shi[1]

[1]College of Mechanical and Electrical Engineering, Harbin Engineering University, Harbin, China

andres@hrbeu.edu.cn, luoani@hrbeu.edu.cn, shidongyan@hrbeu.edu.cn

Abstract. Tensegrities are lightweight structures that, by actuating some or all its components, can become deployable systems that can expand from an initial package to a full-size. This characteristic gives them a great advantage in aerospace applications. In this paper, we present a formulation for the coordinates of tensegrity single units whose only variables are the number of bars and the height of the structure, allowing to know the geometrical configuration of the unit for any given height. We prove that the set of these stable configurations forms a stable path along which a tensegrity unit can be deployed. Numerical simulations using MATLAB, and real models, are used to verify the strategy.

1. Introduction

Tensegrity structures are a special kind of spatial system, com-posed of bars and strings. Because of the way these members are arranged, the internal forces are distributed only along the axes of them in the form of tension in the strings, and compression in the bars. Thanks to this arrangement, other forces such as torsion or torque are not present. This condition complements being lightweight, deployable and the possibility of applying control methods as the main characteristics of the tensegrity structures. Their introduction to the world was in the shape of sculptures; however, in recent years, the engineering has found applications for them in the form of bridges, antennas, masts and robots. The aerospace engineering has taken advantage of the deployability of these structures to develop tensegrity antennas and masts that can be packed into smaller packages than conventional structures, thus reducing costs and weight.

On this topic, Tibert[1] described several applications of deployable tensegrities, Zolesi et al. [2] developed a large deployable antenna reflector based on a double layer tensegrity structure, Sultan and Skelton[3] presented a deployment strategy for tensegrities based on an equilibrium manifold of configurations, Pinaud et al.[4] explored the deployment capabilities of class 2 tensegrity structures by designing and building a prototype model, Rhode-Barbarigos[5] proposed a stochastic search algorithm as the deployment strategy for a tensegrity bridge.

Recently, Sultan[6] used the infinitesimal mechanisms to find paths that produce no loss of energy due to kinematic damping, Porta and Hernández-Juan[7] used differential geometry to propose a quasi-static motion path planning method, Veuve et al.[8] used a stochastic algorithm to develop a bio-inspired learning strategy. However, these approaches are based on advanced mathematical models that de-mand high computational resources. This makes them applicable mostly to multistage, asymmetric and complex tensegrity structures, that can't be modeled for obtaining an analytical solution.

Concerning the equilibrium of single units, Motro[9] demonstrated that the rotation angle between top and bottom planes of a simplex depends only on the number of compressive members; Skelton and De Oliveira[10] expanded this result to derive relationships for the internal force density of the different components of a tensegrity unit that ensure a stable configuration. These relationships, a function of the radius and number of bars of the structure, are widely used in the literature for the form finding of static tensegrities.

In this paper we propose a multi-stable deployment strategy for unitary tensegrity structures. We prove that, making use of the static equilibrium conditions of a tensegrity unit, it is possible to obtain a deployment path along which the structure is stable for any desired height. The nodal coordinates of the structure are reformulated as a function of the height, keeping the length of the bars constant. With the rotation angle between planes as the restriction, the coordinates for any desired height can be obtained. This method assumes that the nodes travel only along stable positions, making unnecessary the use of dynamic equations and reducing the computational cost.

2. Tensegrity Unit

Complex tensegrity systems can be modeled as the composition of several single units, each unit with n number of nodes, b bars and s strings. In a 3-dimensional structure, the simplest composition has 3 bars and 9 strings. Tensegrity units can be modelled as cylinders of radius r, but not every arrangement gives a prestressable configuration; it is necessary then to obtain the range of values that allows the structure to be in equilibrium.

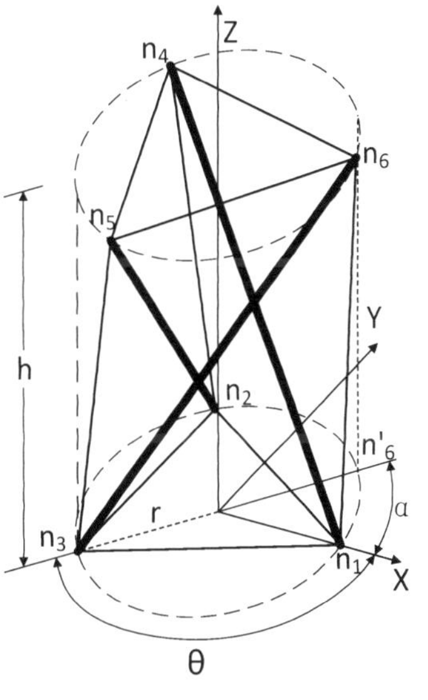

Figure 1. Geometrical parameters of a tensegrity unit.

A tensegrity unit can be defined as two polygons located on parallel planes whose nodes are interconnected. The bottom and top planes have nb nodes each one; the nodes on each plane are connected to their corresponding adjacent nodes by a string, as can be observed in figure 1. Another string connects a bottom node to a top node, located in a position rotated an angle α along Z-axis. A compressive element connects a bottom node n_i with a corresponding top node n_{i+1}, rotated an angle $(\theta+\alpha)$, where θ is $2\pi/b$. It has been demonstrated by Motro [9] and Skelton and De Oliveira [10] that tensegrity simplex are stable only when α is $\pi/2$- π/b. In this paper we only consider the structures known as class 1 tensegrities [10], described as systems where not two or more bars coincide at any node; in other words, bars are only connected by tensile members.

It is straightforward to define the location of the nodes using polar coordinates, as a function of the polygonal angle θ, the twist angle α, the height h and the radius r. In a generic way, the bottom nodes n_b of the structure can be described as

$$n_{bottom} = \begin{bmatrix} r\cos\left(\left(n_i - 1\right)\theta\right) \\ r\sin\left(\left(n_i - 1\right)\theta\right) \\ 0 \end{bmatrix} \tag{1}$$

Using the same notation, and introducing the parameters α and h, the top nodes n_{ti} are described as

$$n_{top} = \begin{bmatrix} r\cos\left(\left(n_i - 1\right)\theta + \alpha\right) \\ r\sin\left(\left(n_i - 1\right)\theta + \alpha\right) \\ h \end{bmatrix} \tag{2}$$

3. Deployment Strategy

The deployment process consists of rising the structure from an initial almost flat configuration to the desired final position. It is assumed that both the collapsed and the deployed arrangements are symmetrical prestressable configurations; it is also assumed that all the tendons are maintained under tension during the process, and the bars don't come in contact between each other. There are no external forces or torque acting throughout the deployment.

During the deployment process, it is desirable to control the sequence in function of the length, or height h, of the structure. For this reason, it is useful to rewrite the dependent parameters of equation (1) and equation (2) in terms of the independent parameter h. Tendon control is chosen over bar control, because the task of changing the active length of strings can be easily accomplished by motors attached at the end of the bars, while modifying the length of bars requires more complex equipment. In this way, the length L of the bar is fixed as constant. A new variable is introduced, the polar angle φ, as the right angle between the Z-axis and any of the bars, as detailed in figure 2, and given by the relation

$$\varphi = \cos^{-1}\left(\frac{h}{L}\right) \tag{3}$$

Through the deployment course, the lengths of the top and bottom strings change symmetrically, resulting in a variation of the radius r that describes the cylinder. The radius of the circumscribed circle is then described as a function of the polar angle φ as

$$r = \frac{1}{2}L\sin\left(\varphi\right) \tag{4}$$

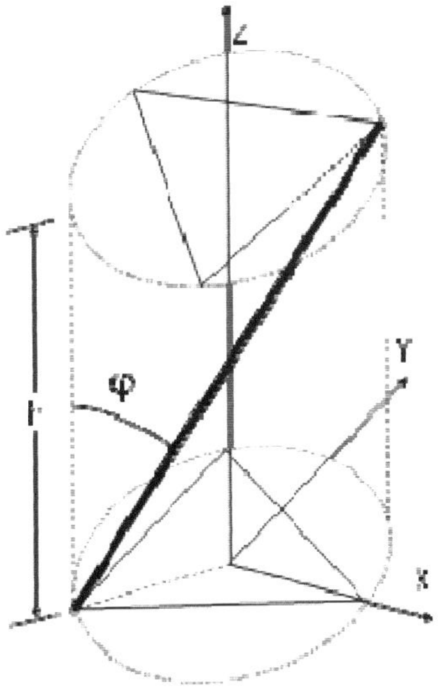

Figure 2. Geometrical parameters for the proposed deployment strategy.

If the relationship of equation (4) is input into the expressions in equation (1) and equation (2), the bottom nodal coordinates become

$$n_{bottom} = \begin{bmatrix} \frac{L}{2}\sin(\varphi)\cos((n_i - 1)\theta) \\ \frac{L}{2}\sin(\varphi)\sin((n_i - 1)\theta) \\ 0 \end{bmatrix} \tag{5}$$

and the coordinates corresponding to the top nodes are also transformed into

$$n_{top} = \begin{bmatrix} \frac{L}{2}\sin(\varphi)\cos((n_i - 1)\theta + \alpha) \\ \frac{L}{2}\sin(\varphi)\sin((n_i - 1)\theta + \alpha) \\ h \end{bmatrix} \tag{6}$$

The polygonal angle θ is a constant geometric parameter for a given number of bars b; it has been proven[9] that α is also a constant parameter that depends only on the number of bars, and L is a design constant representing the length of the compressive elements. Under these conditions, equation (5) and equation (6) give the nodal coordinates as a function where the only variable is the desired h, given that b is defined a priori. For such a configuration, the lengths of the horizontal S_l and vertical strings S_v are given by

$$S_1 = \sqrt{L^2 \sin^2(\varphi)\sin^2\left(\frac{\theta}{2}\right)} \tag{7}$$

IOP Conf. Series: Materials Science and Engineering **649** (2019) 012015 doi:10.1088/1757-899X/649/1/012015

$$S_v = \sqrt{L^2 \sin^2\left(\frac{\alpha}{2}\right)\sin^2(\varphi) + h^2} \tag{8}$$

4. Numerical Example

To validate the results, the deployment courses of a tensegrity structures is presented. Firstly, the software MATLAB is used for simulating the deployment of a single unit, including the nodal coordinates and the force densities. The analytical solution of the problem has been done for a structure with any number b of bars, so the simplest tensegrity structure, composed of 3 bars, is chosen. Next, real models of the positions calculated here are shown to verify the results.

The 3-bar single unit has been widely studied by Skelton and de Oliveira [10] and Zhang, Guest and Ohsaki [11], among several other authors. For this example, a bar of length L=240 mm and a final structural height h of 200 mm are chosen as design parameters.

An arbitrary initial height h of 20 mm is determined, and the deployment course is divided into 4 steps: initial position, two intermediate positions and the final height. The total deployment distance is of 180 mm, divided into steps of 60 units each one. The nodal coordinates for each step are obtained using the equations derived in equation (5) and equation (6). The values of the coordinates for the initial position are shown in table 1, and the graphic of this configuration appears in figure 3(a).

Table 1. Coordinates for the configuration of h=20 mm.

AXIS	NODES [mm]					
	1	2	3	4	5	6
x	119.6	-59.8	-59.8	-103.	0.0	103.6
y	0.0	103.6	-103.	59.8	-119.	59.8
z	0.0	0.0	0.0	20.0	20.0	20.0

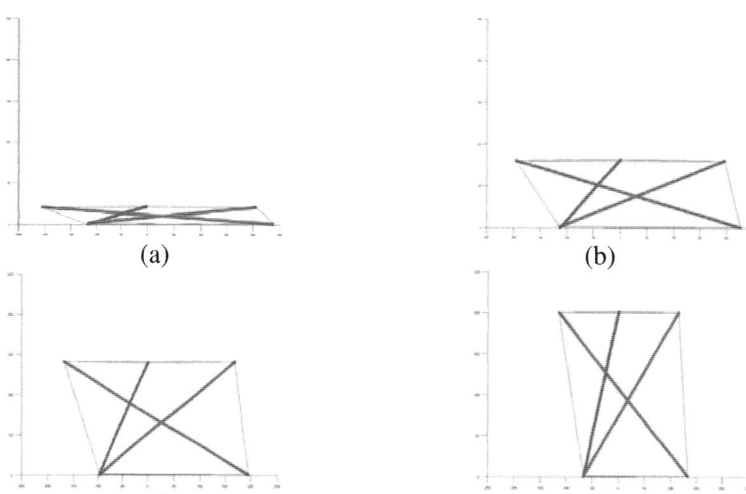

(a) (b) (c) (d)

Figure 3. Initial configuration of the tensegrity unit, (a) h=20 mm, (b) h=80 mm, (c) h=140 mm, (d) h=200 mm.

For this simulation, the initial height is not considered as zero, as a collapsed tensegrity can't have a zero-height because of the thickness of the bars. However, the condition of the minimum height is not studied in this work since it doesn't affect the overall deployment strategy.

Table 2 shows the coordinates for the second position. The value of the x coordinate of node 1 represents the value of the radius r of the cylinder that models the structure. It is noticed how this value is 6.5 mm smaller than the one of the initial configuration; as expected, the radius of the structure is reduced while the height increases. This effect is also noticed from the graphic in figure 3(b).

Table 2. Coordinates for the configuration of *h*=80 mm.

AXIS	NODES [mm]					
	1	2	3	4	5	6
x	113.1	-56.6	-56.6	-97.9	0.0	97.9
y	0,0	97.9	-97.9	56.6	-113.	56.6
z	0.0	0.0	0.0	80.0	80.0	80.0

The coordinates for the third position are also calculated and shown in table 3. In a similar situation as the x coordinate of the node 1, the value of the y coordinate of the node 5 also represents the value of the radius r of the structure. Comparing these values, as seen in figure 3©, allows to confirm that the deployment strategy maintains the overall cylindrical shape throughout the steps.

Table 3. Coordinates for the configuration of *h*=140 mm.

AXIS	NODES [mm]					
	1	2	3	4	5	6
x	97.5	-48.7	-48.7	-84.4	0.0	84.4
y	0.0	84.4	-84.4	48.7	-97.5	48.7
z	0.0	0.0	0.0	140.0	140.0	140.0

The final configuration, as seen in figure 3(d), has the desired height of 200 mm. From the table 4 can be confirmed that the bottom and top radius have the same value. Also, by calculating the distance between nodes 1 and 4, it is confirmed that the length of the bars is unchanged, at a length of 240 mm.

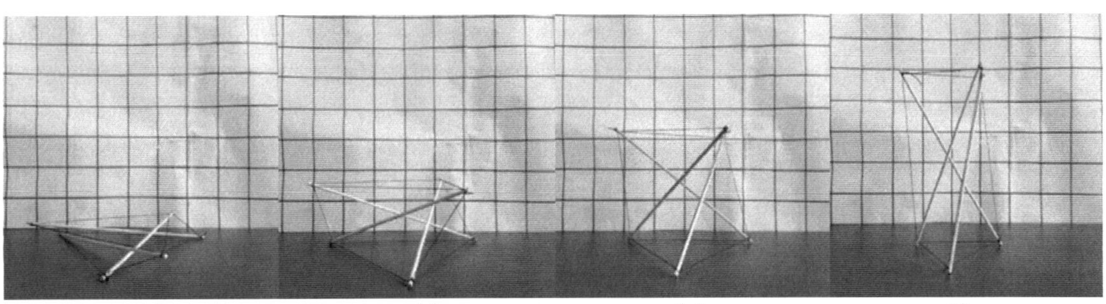

(a) (b) (c) (d)

Figure 4. Initial configuration of the real tensegrity unit.
(a) *h*=20 mm, (b) *h*=80 mm, (c) *h*=140 mm, (d) *h*=200 mm.

Figure 4 shows real models built with the same parameters as the simulated and shown in figure 4. These structures, made of nylon strings and wooden bars, verify that the analytical and numerical results can be applied in real applications. The length of the strings of each structure are calculated from the coordinates in table of the corresponding configuration using equation (7) and equation (8).

Table 4. Coordinates for the configuration of h=200 mm.

AXIS	NODES [mm]					
	1	2	3	4	5	6
x	66.3	-33.2	-33.2	-57.5	0.0	57.5
y	0.0	57.5	-57.5	33.2	-66.3	33.2
z	0.0	0.0	0.0	200.0	200.0	200.0

5. Conclusions

In this paper, a multi-stable deployment strategy for tensegrity structures has been presented. The nodal coordinates as a function of the height of the units were derived, based on the static equilibrium condition for tensegrity units. It was proven that, based on these two sets of information, it is possible to obtain a deployment course along which the tensegrity structure is always stable. Examples of a prismatic single unit, simulated in MATLAB and real models, were given for verification purposes.

The advantage of this deployment strategy is that it avoids the use of dynamic equations, simplifying the algorithm and reducing the computational cost. Further research should consider the deployment of multi-stage tensegrity structures.

References

[1] Tibert G 2002 Deployable tensegrity structures for space applications, PhD Dissertation, Royal Institute of Technology, Stockholm, 2002.
[2] Zolesi V S, Ganga P L, Scolamiero L, Micheletti A, Podio-Guidugli P, Tibert G, et al. 2012 *42nd Int. Conf. Environ. Syst.* On an innovative deployment concept for large space structures, 2553-66.
[3] Sultan C and Skelton R E 2003 *Int. J. Solids Struct.* Deployment of tensegrity structures, **40** 4637-57.
[4] Pinaud J P, Solari S and Skelton R E 2004 *Smart Struct. Mater.* Deployment of a class 2 tensegrity boom, 155-62.
[5] Rhode-Barbarigos L 2012 An active deployable tensegrity structure, PhD Dissertation, École Polytechnique Fédérale de Lausanne, Lausanne, Switzerland.
[6] Sultan C 2014 *Int. J. Solids Struct.* Tensegrity deployment using infinitesimal mechanisms, **51** 3653-68.
[7] Porta J M and Hernandez-Juan S 2016 *Int. J. Solids Struct.* Path planning for active tensegrity structures, **78-79** 47-56.
[8] Veuve N, Sychterz A C and Smith I F C 2017 *Eng. Struct.* Adaptive control of a deployable tensegrity structure, **152** 14-23.
[9] Motro R 2003 Tensegrity: Structural Systems for the Future, Kogan Page Science, London.
[10] Skelton R E and De Oliveira M 2009 Tensegrity Systems, first ed., Springer US, New York.
[11] Zhang J, Guest S D and Ohsaki M 2009 *Int. J. Solids Struct.* Symmetric prismatic tensegrity structures: Part I. Configuration and stability, **46** 1-14.

Acknowledgements

This work was supported by the National Natural Science Foundation of China (NSFC), grant number 51605111, 51675114.

Finite Time Cascade Control Scheme for Tracking Control of an Underwater Exploring Robot

Zhimin Liu[1] and Xiaozhong Zhou[1]

[1]Basic department, Shanghai publishing and printing college, Shanghai 200093, China

qiierrsann@163.com

Abstract: In this paper, we discuss the curve trajectory tracking problem of an underactuated underwater robot in the horizontal plane. We use global diffeomorphism transformation method to transform the sixth-order dynamic tracking error model into a nonlinear cascaded system, by using the cascaded systems' finite time stability theory, along with the back stepping technique to design the controller. Two discontinuous finite time controllers are designed for the two subsystems respectively. In the existing literature, finite time control is used for the first order nonholonomic system, we extended this method to two order nonholonomic system in this paper. The finite time controller can tune not only the gain, but also the fractional power, it is more convenience for engineering implementation. The results show the control laws can achieve the control objective with good performance for the horizontal trajectory tracking control.

1. Introduction

This paper, we discuss the curve trajectory tracking problem of an under-actuated underwater vehicle (UUV) in the horizontal plane. The difficulty for controlling the under-actuated UUV is the limitations of the Brockett necessary condition [1]. To overcome the difficulty, the discontinuous feedback approach and time-varying feedback approach are often used [2]. The mathematical model of an under-actuated UUV [3] moving in a horizontal plane can be described as equation (1):

$$\begin{gathered} m_{11}\dot{u} - m_{22}vr + d_{11}u = \tau_1, \\ m_{22}\dot{v} - m_{11}ur + d_{22}v = 0, \\ m_{33}\dot{r} + (m_{22} - m_{11})uv + d_{33}r = \tau_3, \\ \dot{x} = ucos\varphi - vsin\varphi, \\ \dot{y} = usin\varphi + vcos\varphi, \\ \dot{\varphi} = r, \end{gathered} \tag{1}$$

Where u, v and r represent the velocities of a UUV in surge, sway and yaw in the body-fixed frame respectively; x, y, φ represent the position and orientation of a UUV in the earth-fixed frame. The parameters $m_{ii} > 0$ are the inertia and added mass effects of the UUV. The parameters $d_{ii} > 0$ represent the hydrodynamic damping of the UUV. The available input controls are the thruster force τ_1, and the steering torque τ_3. Due to the absence of actuators in sway direction, the discussed UUV is an under-actuated two order nonholonomic systems.

For under-actuated nonholomic underwater vehicles, many control strategies have been proposed for AUV trajectory tracking controller, such as the local linearization, the back stepping method, adaptive neural network control, the Sliding Mode control method, intelligent control method, the Lyapunov design and redesign, etc. [4, 5].

Note that most of the aforementioned control schemes only consider the asymptotic stability, which means that the systems considered achieve convergence in infinite settling time, or have at best exponential convergence rate to the origin.

Finite time control aims to make the dynamical errors converging to the origin in finite time. Finite time stability means the controlled plant must be finite time convergence and satisfy the Lyapunov stability theorem. In [6], the sufficient condition for finite time stability of nonlinear dynamical systems was given. With the improvement of finite time control theory such as homogeneous theory and back stepping theory [7, 8], the design methods for non-Lipschitz continuous finite time controllers have developed rapidly and many new results for designing finite time controller have been obtained, see for [9-11] and the references therein. Compared with the classical asymptotic design scheme, the finite time controller possesses the property which can steer the dynamical errors converge to the origin in a finite time, furthermore, the finite time controller usually demonstrates higher tracking accuracy robustness against uncertainties and better disturbance-rejection performance [12].

The finite time control has attracted the interest of many scholars in recent years, but the research is mainly focused on the first order nonholonomic system, such as the wheeled mobile robots, the velocity along the plane perpendicular to the point of contact between the wheel and the ground is zero. The UUVs belong to two order nonholonomic system, and its control problem is more challenging. We extended the finite time method to two order nonholonomic systems in this paper.

Assume that a feasible reference trajectory $(u_d, v_d, r_d, x_d, y_d, \varphi_d)$ is given, i.e., a trajectory, $u_d, v_d, r_d, x_d, y_d, \varphi_d$ satisfy the same dynamics equations with $u, v, r\, x, y, \varphi$.

We use the same coordinate and input transformations as in [13]. Let e=$[e_1, e_2, e_3, e_4, e_5, e_6]^T = [z_1 - z_{1d}, z_2 - z_{2d}, z_3 - z_{3d}, z_4 - z_{4d}, z_5 - z_{5d}, z_6 - z_{6d}]^T$, we can obtain the tracing error dynamics equations

$$\begin{cases} \dot{e}_1 = -ae_1 - ae_4 + (e_2e_6 + z_{2d}e_6 + z_{6d}e_2) - b(e_5e_6 + z_{5d}e_6 + z_{6d}e_5) \\ \dot{e}_2 = e_4e_6 + z_{4d}e_6 + z_{6d}e_4 \\ \dot{e}_3 = e_6 \\ \dot{e}_4 = F_1 - F_{1d} \\ \dot{e}_5 = c(e_1e_6 + z_{1d}e_6 + z_{6d}e_1 + e_4e_6 + z_{4d}e_6 + z_{6d}e_4) - ce_5 \\ \dot{e}_6 = F_2 - F_{2d} \end{cases} \tag{2}$$

From the analysis of the above, the control objective moving the BYSQ-3 from the initial trajectory to a desired trajectory in the configuration space can be described as the stabilization of $[e_1, e_2, e_3, e_4, e_5, e_6]^T$.

2. Finite Time Controller Design

2.1. Designing the Finite Time Control Laws for e_3, e_6

Let $\alpha(e_6)$ be the virtual input of e_6 and $\alpha(e_6) = -l_3 sig^{\rho}(e_3)$, where $l_3 > 0,\ 0 < \rho < 1$, define

$$s_6 = e_6 - \alpha(e_6) \tag{3}$$

substituting (3) into (2), we have:

$$\begin{cases} \dot{e}_3 = s_6 - l_3 sig^{\rho}(e_3) \\ \dot{s}_6 = F_2 - F_{2\mathrm{d}} - \dot{\alpha}(e_6) \end{cases} \tag{4}$$

Take the following Lyapunov function

$$V_1 = \frac{1}{2}(e_3^2 + s_6^2) \tag{5}$$

Differentiating (5) along the solutions of the (4) satisfies

$$\dot{V}_1 = e_3\dot{e}_3 + s_6\dot{s}_6$$
$$= -l_3 e_3 sig^{\rho}(e_3) + e_3 s_6 + s_6(F_2 - F_{2\mathrm{d}} - \dot{\alpha}(e_6))$$

If we design

$$F_2 = F_{2\mathrm{d}} + \dot{\alpha}(e_6) - e_3 - l_6 sig^{\rho}(e_6),\ l_6 > 0, \tag{6}$$

Then

$$\dot{V}_1 = -l_3 e_3 sig^{\rho}(e_3) + e_3 s_6 + s_6(F_2 - F_{2\mathrm{d}} - \dot{\alpha}(e_6))$$
$$\leq -h(\frac{1}{2}e_3^2 + \frac{1}{2}s_6^2)^{\frac{1+\rho}{2}}$$
$$= -hV_1^{\omega},$$

where $h = 2^{\omega}min(l_3, l_6)$,$0 < \omega = \frac{1+\rho}{2} < 1$.

From the theorem in [14], the control law moves the states e_3, e_6 to zero in finite time t_1. at the end of the first stage we have $e_3 = e_6 = 0$

Then the system (2) is convert into

$$\begin{pmatrix} \dot{e}_1 \\ \dot{e}_2 \\ \dot{e}_4 \\ \dot{e}_5 \end{pmatrix} = \begin{pmatrix} -ae_1 - ae_4 + e_2 z_{6d} - be_5 z_{6d} \\ e_4 z_{6d} \\ F_1 - F_{1\mathrm{d}} \\ -ce_5 + c(e_4 z_{6d} + e_1 z_{6d}) \end{pmatrix} \tag{7}$$

We design the virtual control input of e_4 to steer the state variables to origin in finite time. Considering the subsystem (7), let $\alpha(e_4)$ be the virtual control input of e_4 and define

$$s_4 = e_4 - \alpha(e_4), \tag{8}$$

combining (7) and (8), we can obtain

$$\begin{pmatrix} \dot{e}_1 \\ \dot{e}_2 \\ \dot{s}_4 \\ \dot{e}_5 \end{pmatrix} = \begin{pmatrix} -ae_1 - a(s_4 + \alpha(e_4)) + e_2 z_{6d} - be_5 z_{6d} \\ (s_4 + \alpha(e_4))z_{6d} \\ F_1 - F_{1d} - \dot{\alpha}(e_4) \\ -ce_5 + c((s_4 + \alpha(e_4))z_{6d} + e_1 z_{6d}) \end{pmatrix} \tag{9}$$

Take the following Lyapunov function

$$V_2 = \frac{1}{2}(e_1^2 + e_2^2 + s_4^2 + e_5^2) \tag{10}$$

Differentiating (10) along the solutions of the (9) satisfies

$\dot{V}_2 = e_1\dot{e}_1 + e_2\dot{e}_2 + s_4\dot{s}_4 + e_5\dot{e}_5 =$
$-ae_1^2 - ce_5^2 + \alpha(e_4)(-ae_1 + e_2 z_{6d} + ce_5 z_{6d}) +$
$s_4\big(-ae_1 + e_2 z_{6d} + ce_5 z_{6d} + F_1 - F_{1d} - \dot{\alpha}(e_4)\big) + e_1 e_2 z_{6d} + (c-b)e_1 e_5 z_{6d}.$

If we design $\alpha(e_4)$ as

$\alpha(e_4) = l_1 sig^{\sigma}(e_1) - l_2 z_{6d} sig^{\sigma}(e_2) - l_5 z_{6d} sig^{\sigma}(e_5)$,

where l_1, l_2, l_5 are positive real numbers, σ is a real number and $0 < \sigma < 1$,

and we design F_1 as

$F_1 = F_{1d} + \dot{\alpha}(e_4) + ae_1 - e_2 z_{6d} - ce_5 z_{6d} - l_4 sig^{\sigma}(s_4)$,

where l_4 is a positive real number,

then, we can obtain

$\dot{V}_2 = e_1\dot{e}_1 + e_2\dot{e}_2 + s_4\dot{s}_4 + e_5\dot{e}_5 =$
$-ae_1^2 - ce_5^2 + \alpha(e_4)(-ae_1 + e_2 z_{6d} + ce_5 z_{6d}) +$
$s_4\big(-ae_1 + e_2 z_{6d} + ce_5 z_{6d} + F_1 - F_{1d} - \dot{\alpha}(e_4)\big) + e_1 e_2 z_{6d} + (c-b)e_1 e_5 z_{6d}$

where

$\Delta = e_1 e_2 z_{6d} + (c-b)e_1 e_5 z_{6d} - ae_1\big(-l_2 z_{6d} sig^{\sigma}(e_2) - l_5 z_{6d} sig^{\sigma}(e_5)\big)$
$+ e_2 z_{6d}\big(l_1 sig^{\sigma}(e_1) - l_5 z_{6d} sig^{\sigma}(e_5)\big) + ce_5 z_{6d}(l_1 sig^{\sigma}(e_1) - l_2 z_{6d} sig^{\sigma}(e_2))$,
$l = 2^{\tau} l', l' = \min\{al_1, l_2 z_{6d}^2, l_4, cl_5 z_{6d}^2\}$
$0 < \tau = (1+\sigma)/2 < 1$.

If we choose proper real numbers $l_i (i = 1,2,\cdots,5)$ such that the following inequality can hold:

$$l > p + \Delta / V_2^{\tau}, p > 0, \tag{11}$$

then, we can obtain

$$\dot{V}_2 \leq -pV_2^{\tau}$$

Then from the theorem in [14], we know the control law moves the states e_1, e_2, e_4, e_5 to zero in finite time t_2

3. Simulations

The hydrodynamic coefficients are as follows:

$m_{11} = 45$ kg , $m_{22} = 46$kg, $m_{33} = 45.5$ kg,
$d_{11} = 0.4$kg/s, $d_{22} = 0.2$kg/s, $d_{33} = 0.001$kgm^2/s,

The desired trajectory is chosen as a circle given by

$x_d = 5cost, y_d = 5sint, \varphi_d = t + \pi/2,\ u_d = 3,\ v_d = 0,\ r_d = 1.$

The initial conditions are

$x(0) = 5\text{m}, y(0) = 5\text{m}, \varphi(0) = pi/2, u(0) = 0\text{m/s}, v(0) = 0\text{m/s}, r(0) = 0\,\text{rad/s}.$

Thus, we get a group of control parameters as

$l_1 = 2.4, l_2 = 5.8, l_4 = 3.2, l_5 = 1.8$,
The other parameters are selected as
$k_1 = 2, k_2 = 8, k_3 = 3, k_4 = 0.8, k_5 = 1.2, k_6 = 0.8, k_7 = 1 \quad \alpha = 0.8, \rho = 0.6$.
The simulation results are shown in figures 1-6.

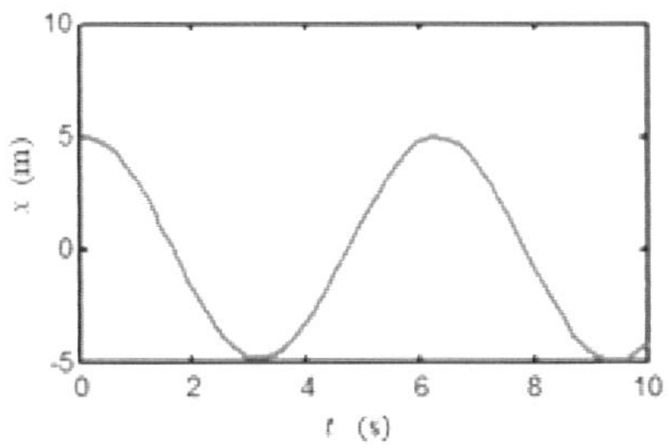

Figure 1. Trace of x.

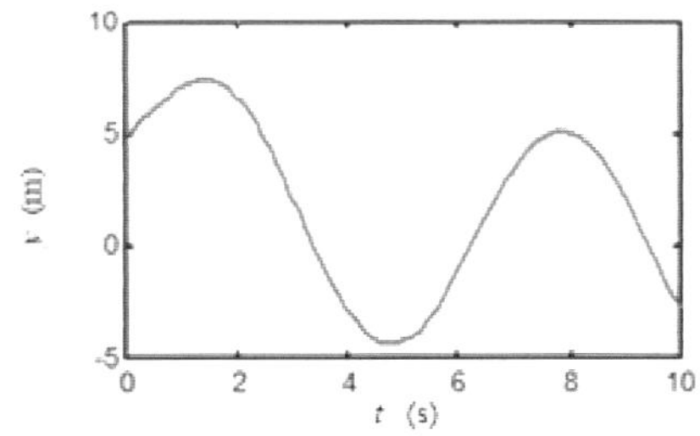

Figure 2. Trace of y.

The AUV moves from stationary state. Its initial tracking errors are $|x_e| = 0$ m, $|y_e| = 5$ m, figures 1, 2 show the time responses of the position and velocity of the AUV. From these simulation results, the error x_e is less than 0.10 m and y_e is less than 0.07 m,.

In figure 3, the simulation trajectory, under the action of the finite time tracking controller is demonstrated. It can be seen obviously that the robot can accomplish the tracking mission accurately in finite time.

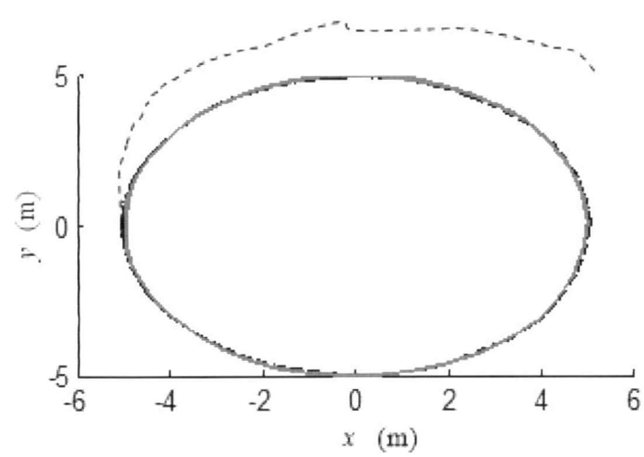

Figure 3. Simulation path of BYSQ-3.

From figure 3, it can be seen that the sway displacement, surge displacement converges to desired trajectory infinite time. From figure 3, it can be seen that the angle increased uniformly, which conform to the situation of tracking circular trajectories.

References

[1] Wadoo S A and Kachroo P 2010 Autonomous underwater vehicles: modeling, control design, and simulation. CRC Press.

[2] Do K D and Pan J 2009 Control of Ships and Underwater Vehicles, Advances in Industrial Control.

[3] Aguiar A P and Hespanha J P 2007 *IEEE Transac. Automatic Control* Logic-based switching control for trajectory-tracking and path-following of underactuated autonomous vehicles with parametric modeling uncertainty, **52(8)** 1362-79.

[4] Du J and Guo C 2004 *2004 American control conf.* Nonlinear adaptive ship course tracking control based on backstepping and nussbaum gain, 3845-50.

[5] Holzhü T 1997 *IEE Proceedings* LQG approach for the high-precision track control of ships. Control Theory and Applications, **144(2)** 121-7.
[6] Haimo V T 1986 *SIAM J. Control Optim.* Finite time controllers. **24(4)** 760-70.
[7] Bhat S P and Bernstein D S 1998 *IEEE Transac. Automatic Control* Continuous finite-time stabilization of the translational and rotational double integrators, **43(5)** 678-82.
[8] Bhat S P and Bernstein D S 2005 *Mathemat. Control Signals Syst.* Geometric homogeneity with applications to finite-time stability, **17(2)** 101-27.
[9] Huang X, Lin W and Yang B 2005 *Automatica* Global finite-time stabilization of a class of uncertain nonlinear systems, **41(5)** 881-8.
[10] Zhai J 2014 *Circuits Syst. Signal Process* Finite-time output feedback stabilization for stochastic high-order nonlinear systems, **33(12)** 3809-37.
[11] Zhang X, Feng G and Sun Y 2012 *Automatica* Finite-time stabilization by state feedback control for a class of time-varying nonlinear systems, **48(3)** 499-504.
[12] Du H B, Li S H and Qian C J 2011 *IEEE Transac. Automatic Control* Finite-time attitude tracking control of space craft with applicationto attitude synchronization, **56(11)** 2711-17.
[13] Ma B L 2009 *Syst. Control Lett.* Global K-exponential asymptotic stabilization of underactuated surface vessels, **58(3)** 194-201.
[14] Li S and Tian Y P 2007 *Int. J. Control* Finite-time stability of cascaded time-varying systems, **80(4)** 646-57.

Hydrodynamic Analysis and Structural Optimization of an Underwater Robot

Zhimin Liu[1] and Xiaozhong Zhou[1]

[1]Basic department, Shanghai publishing and printing college, Shanghai 200093, China

qiierrsann@163.com

Abstract. BYSQ-2 is an amphibious spherical exploring robot which can move flexibly in water and on land, it has good water pressure resistance and can perform a rotational motion with a zero degree turn radius. But the actual experiment results show the shortcoming of the robot is the low maximum surge speed. The main objective in this paper is to improve the maximum surge speed and alleviate the heavy pendulum's oscillation. Firstly, the hydrodynamic coupling under different structural parameters and propeller working states is analyzed by using CFD method, the influence of structure parameters and propeller rotation speed on thrust and water resistance are summarized, based on the optimal structural parameters, the new generation robot with better performance is designed. The simulation and experiment results show the validity of the new design scheme.

1. Introduction

Hydrodynamic characteristics analyses are important for achieving necessary maneuverability, the existed approaches for acquire hydrodynamic characteristics often including scaled or full-scale experiments, empirical formula approximations, system identification and computational dynamic approaches [1, 2], Experimental method is in high-accuracy but can only obtain limited measure results with expensive cost, empirical formula approximations method requires deeper knowledge and experiences while it is not comprehensive [3]. CFD software such as SHIPMO^{TM}, ANSYSCFX^{TM}, ANSYS-FLUNT^{TM} and ANSYS-AQWA^{TM} are capable to predict hydrodynamic parameters for a complex-shaped AUV with very low cost and it can supply engineers with accurate information for potential vehicle designs and control algorithms [4-8]. CFD technique is mainly used to analyze the streamline underwater robot, the applications on spherical underwater robot is still relatively few.

2. Mechanical Structure of BYSQ-2 Prototype and Experiment

BYSQ-2 is an amphibious spherical roving robot which can move flexibly in water and on land, it has a fiber glass external hull, the internal structure of BYSQ-2 mainly including a single propeller, the heavy pendulums and a flywheel. The catheter is fixed connected with the spherical shell, the propeller is mounted in the middle of the catheter, the sleeve (or long axis) is mounted the outer wall of the catheter and it can rotate around the catheter, the short axis mechanism and the flywheel rotation axis is fixed to sleeve and perpendicular to the sleeve. The robot regulate its attitude by using the heavy pendulums rotating around the long or short axis

In addition to the above advantages, the spherical structure has a drawback, that is, the water resistance is larger than the streamlined structure which results the low velocity in moving forward. The movement resistance of underwater objects includes linear damping term and nonlinear damping term, due to the low speed of the robot, we neglect the linear damping term and the Coriolis forces, based on Newton's second law:

$$F=ma \tag{1}$$

the surge dynamic equation can be expressed as:

$$(m+\lambda)\dot{v}=T-F_D \tag{2}$$

Where m denotes the mass of the robot, λ denotes the added mass of water and can be calculated by using radiation/diffraction program MCC and WAMITTM, T represents the Propeller's thrust, v is the surge velocity, F_D is the movement resistance of the robot and can be expressed as:

$$F_D=\left(0.632+\frac{4.8}{\sqrt{\mathrm{Re}}}\right)^2\rho\frac{v^2}{2}A \tag{3}$$

Combing the equations (1)-(3), we can obtain:

$$(m+\lambda)\dot{v}+\left(0.632+\frac{4.8}{\sqrt{\mathrm{Re}}}\right)^2\rho\frac{v^2}{2}A-K_T\rho n^2 D_p^{\ 4}=0 \tag{4}$$

From equation (4), the maximum velocity can be obtained as:

$$v_{\max}=\frac{2K_T\rho n^2 D_p^{\ 4}}{\sqrt{\rho A}\left(0.632+\frac{4.8}{\sqrt{\mathrm{Re}}}\right)} \tag{5}$$

In order to verify the maximum velocity of the theory calculation, the real world experiments have been conducted in the calm swimming pool at BUPT, after the robot reached the maximum velocity, the time period was recorded for the robot moving through 1m distance (see figure 3), The experiment was repeated five times, and the average maximum velocity was calculated based on the experiment results. The experiment recorder is described in table 1.

Table 1. Test data of maximum velocity.

Cycle	1	2	3	4	5
Time (s)	1.53	1.54	1.52	1.54	1.53
Test results(m/s)	0.655	0.654	0.656	0.644	0.655
Computational result(m/s)			0.681		

3. Resistance and Thrust Analysis under Different Structural Parameters

The resistance of shell and the thrust of the propeller are the main factor which can affect the robot's surge velocity. Moreover, the propeller is mounted in the middle of the catheter, the propulsion ability is closely related to structural parameters of the spherical shell in addition to the structure of itself and the propeller's rotational speed, hence the structural parameters, resistance and the propeller's thrust couple and interact with each other. In previous work, we had assumed that the robot was an absolute spherical, so the hydrodynamic characteristics were the same for all DOFs. But the planar mechanic machine (PMM) test shown the add mass in surge direction and sway direction was obviously different. To ensure the accuracy of the model and to simplify the analysis of the coupling hydrodynamic coefficients, the sphere-pipe model is established, the bolts were ignored to reduce the complexity of the mesh and enhance the mesh quality. The BYSQ-2 was modelled in a cube flow field with 2 m side length, the mesh density was increased around the robot and three boundary layers were used to obtain satisfactory results and reduce the amount of time required.

We carry out the simulations Under the conditions of v=1m/s, d = 0.14m, n= 600 rpm and shell diameter D changes from 0.14 m to 0.5 m, the simulation results are shown as in figure 1.

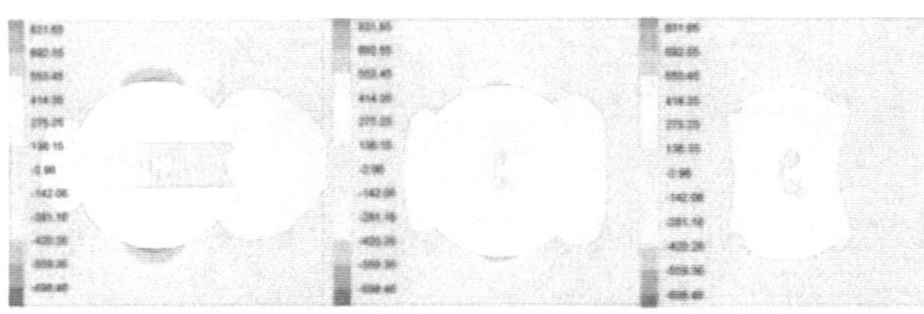

Pressure change under different spherical shell diameter

Figure 1. Influence of different spherical diameter.

From figure 1, it can be seen the pressure in the front is increasing with *D* increasing and it can be seen that with the propeller's diameter *D* increasing, the move forward resistance is increased obviously, while the propeller's thrust change little in the process, so, minimizing the size of the spherical shell is another method to improve the robot's velocity.

Figure 1 shows that the thrust of the propeller substantially contains unchanged, so if the propeller's rotation speed *n* keeps constant, the thrust of the propeller is not affected by the changing of the shell diameter and the catheter diameter, the move forward resistance of the robot can be reduced by increasing the catheter diameter *d* and decreasing the shell diameter *D*. However, the catheter diameter can't be too large, and the sphere shell diameter *D* of the robot can't also be too small, because the sleeve is mounted the outer wall of the catheter and it need to rotate around the catheter and the heavy pendulum and the control circuit occupy the internal space of the robot. Therefore, the optimal parameters are chosen 0.14 and 0.35 for the catheter diameter *d* and the shell diameter *D*, respectively. in order to verify the performance of the optimal parameters, the simulation is carried out under the conditions of *v*=1m/s, and propeller's rotation speed *n* change from 0 to 2000 rpm, the simulation results of the optimal parameters are shown in figure 2.

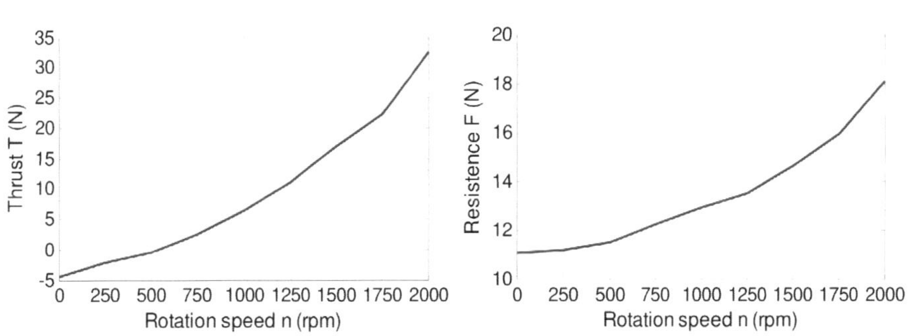

Figure 2. Influence of different propeller rotation speed.

As can be seen from figure 2, when the propeller's rotation speed increasing, the water pressure in front of the robot keeps almost no change, but gradually becomes smaller in the rear of the robot, the water pressure also gradually becomes smaller in the front of propeller.

In order to further clarify the hydrodynamic coupling under different structural parameters and propeller working states, we have already conducted a lot of simulation analysis, the results can be seen in figure 3.

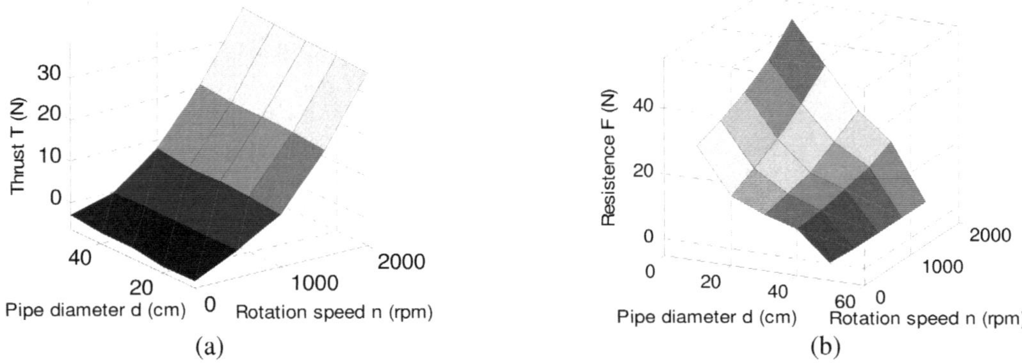

Figure 3. Hydrodynamic Analysis of different key parameters.

Figure 3 Analysis the key parameters which can affect the resistance and propeller's force including the catheter diameter *d*, the spherical shell diameter *D* and propeller rotation speed *n* in detailed. From figure 3, it can be seen that the propeller's rotation speed is the main factor which influence the propeller's thrust; From figure 3, the move forward resistance of the robot is affected by all of the key parameters, simultaneous it is affected by the propeller rotation speed *n*; the real world experiments have been conducted in the same conditions as in section 2, from the experimental data, we calculated that the robot's maximum speed is 1.2 m/s.

4. Conclusion

To solve the problem of low maximum moving velocity, the coupling hydrodynamic characteristic under different key parameters is analyzed. Through simulation and experiment, we found that it is effective for improving the maximum moving velocity by increasing the catheter diameter and decreasing the spherical shell diameter, with the propeller rotation speed increasing, water resistance, the propeller's thrust, torque and output power increase apparently. According to the above analysis, combined with the internal structure of the robot, we determine the optimal structure parameters and develop a new generation underwater spherical roving robot. The real world experiment show the new generation robot's maximum speed is about 1.2 m/s.

References

[1] Vervoort J H A M 2009 Modeling and Control of an Unmanned Underwater Vehicle.
[2] Coe R G 2013 Improved Underwater Vehicle Control and Maneuvering Analysis with Computational Fluid Dynamics Simulations, Virginia Tech.
[3] Zhang H, Xu Y R and Cai H P 2010 *J. Marine. Sci. Appl.* Using CFD Software to Calculate Hydrodynamic Coefficients, **9** 149-55.
[4] Yang R, Clement B, Mansour A, et al. 2015 *J. Intel. Robot. Syst.* Modeling of a Complex-Shaped Underwater Vehicle for Robust Control Scheme, **80(3-4)** 491-506.
[5] Ferziger J H and Peric M 1996 Computational method for fluid dynamics, Springer, Berlin, 28-50.
[6] Sarkar T, Sayer P G and Fraser S M 1997 *Int. J. Numerical Methods Fluids* A study of autonomous underwater vehicle hull forms using computational fluid dynamics, **25(11)** 1301-13.

[7] Tyagi A and Sen D 2006 *Ocean Eng.* Calculation of transverse hydrodynamic coefficients using computational fluid dynamic approach, **33(5)** 798-809.

[8] Wilson R, Paterson E and Stern F 2006 Unsteady RANS CFD method for naval combatant in waves, Proceedings of the 22nd ONR Symposium on Naval Hydrodynamics, Washington DC, National Academy Press, 532-49.

IOP Conf. Series: Materials Science and Engineering **649** (2019) 012018 doi:10.1088/1757-899X/649/1/012018

eMineProve: Educational Data Mining for Predicting Performance Improvement Using Classification Method

Jovel T. Rosado[1], Angelica P. Payne[2] and Corazon B. Rebong[3]

[1]University of Perpetual Help System Laguna, Philippines
[2]Olivarez College, Tagaytay, Philippines
[3]Colegio de San Juan de Letran, Calamba, Philippines

rosado.jovel@uphsl.edu.ph, angiepayne5@gmail.com, corazonbrebong@gmail.com

Abstract. Today, Data mining has become a universal tool for converting data into useful information and knowledge. Student's academic performance is a crucial factor in building their future [3]. In this study, the classification method is selected to be applied on the students' data. Classification is a form of data analysis that extracts models describing data classes [4]. The data is collected from the Basic Education Department of the University of Perpetual Help System Laguna. The raw data set is a collection of 4, 250 data accumulated over two academic years regarding the basic information of the Grade 7 Junior High School students. The data mining technique applied was classification using Naïve Bayes. The data mining tool used to process the data into useful knowledge is RapidMiner. The main objective of this study is to predict the performance improvement of Grade 7 Junior High School students for A.Y. 2016-2017 and 2017-2018 using classification. It specifically aims to identify the general average of the Grade 7 Junior High School students when grouped according to gender, identify who perform better between male and female, identify the subject on which the students excel most, identify the subject on which students have difficulty, identify who performs best when group according to last school attended, identify the academic performance of the students based on their parent's occupation, provide a predictive analysis of data to help the decision makers create a marketing strategy for those schools where only few students enrolled. Naïve Bayes produces accuracy 92.37% that shows it is possible to obtain a good prediction model based on the academic performance of the students. Based on the data for marketing strategy, the model has an accuracy of 30.97%.

1. Introduction

As our society is growing in terms of population, technology advancements, literary rate and global competitiveness, education also is taking a leap as business systems do. Today, Data mining has become a universal tool for converting data into useful information and knowledge. Data Mining (DM) also known as Knowledge Discovery from Databases (KDD), is the area of discovering unique and potentially useful information from the large amount of data [1].

Educational Data Mining (EDM) is the application of Data Mining techniques on educational data. To analyze data and to resolve educational research issues is the main objective of EDM. It deals with developing new approaches to explore the educational data, and using it to better understand student learning environment [2].

Student's academic performance is a crucial factor in building their future [3]. In this study, the classification method is selected to be applied on the students' data. Classification is a form of data analysis that extracts models describing data classes [4].

The main objective of this study is to predict the performance improvement of Grade 7 Junior High School students for A.Y. 2016-2017 and 2017-2018 using classification. It specifically aims to identify the general average of the Grade 7 Junior High School students when grouped according to gender, identify who perform better between male and female, identify the subject in which the students excel most, identify the subject in which students have difficulty, identify who performs best when group according to last school attended, identify the academic performance of the students based on their parent's occupation, provide a predictive analysis of data to help the decision makers create a marketing strategy for those schools where only few students enrolled.

2. Related Works

A number of researchers have published work that studied various EDM techniques used to better understand and enhance student performance. Applied association rules data mining techniques on learner data to evaluate knowledge about learners' academic performance [5].

K. Shanmuga Priya and A. V. Senthil Kumar [6] applied a Classification Technique in Data Mining to improve the student's performance and help to achieve the goal by extracting the discovery of knowledge from the end semester mark.

Monika Goyal and Rajan Vohra [7] applied data mining techniques to improve the efficiency of higher education institution. If data mining techniques such as clustering, decision tree and association are applied to higher education processes, it would help to improve students' performance, their life cycle management, selection of courses, to measure their retention rate and the grant fund management of an institution

Cortez and Silva [8] attempted to predict failure in the two core classes (Mathematics and Portuguese) of two secondary school students from the Alentejo region of Portugal by utilizing 29 predictive variables.

Ramaswami et.al.[9], developed a predictive data mining model for students' performance so as to identify the slow learners and study the influence of the dominant factors on their academic performance.

3. Methodology

3.1. Data Preprocessing Phase

Data of 4,250 from the Grade 7 Junior High School students of the Basic Education of the University of Perpetual Help System Laguna from A.Y. 2016-2017, 2017-2018. There are total of 250 students of Grade 7 Junior High School for the two succeeding years. The data were consisting of the gender of the students, originating schools, address, parents' occupation, final ratings in all subject areas and their final average.

The dataset has to go through a preprocessing phase to clean the data from different errors and remove non-important and redundant attributes. The required data for evaluating students' performance are scattered over many tables; thus, this phase starts with attribute selection and merging multiple tables into a single table that contains the most important attributes for the required evaluation.

The last step in this phase is the transformation of the dataset into a suitable format for the mining algorithms.

3.2. Data Mining Phase

In this phase, classification technique is used using Naïve Bayes are applied using RapidMiner software. RapidMiner tool is used for exploration, statistical analysis and mining the students' data. Among the algorithms, Naïve Bayes as classification technique will be used. Naïve Bayes algorithm is a simple probabilistic classifier that calculates a set of probabilities by counting the frequency and combinations of values in a given data set. The data classification process involves learning and classification. In learning the training data are analyzing by classification algorithm. In classification test data are used to estimate the accuracy of the classification rules. [10]

Likelihood Class Prior Probability

$$P(c|x) = \frac{P(x|c)P(c)}{P(x)}$$

Posterior Probability Predictor Prior Probability

$$P(c|X) = P(x_1|c) \times P(x_2|c) \times \cdots \times P(x_n|c) \times P(c) \qquad (1)$$

Equation (1) shows Bayes Theorem which provides a way of calculating posterior probability P(clx) from P(c), P(x), and P(xlc). Here;

- P(clx): posterior probability of class (c, target) given predictor (x, attributes). This represents the probability of c being true, provided x is true.
- P(c): is the prior probability of class. This is the observed probability of class out of all the
- observations.
- P(xlc): is the likelihood which is the probability of predictor-given class. This represents the probability of x being true, provided x is true.
- P(x): is the prior probability of operator. This is the observed probability of predictor out of all the observations.

Table 1 presented the variables used in data mining that include the gender, last school attended, subjects, and parents' occupation, grade per subject, general average, academic performance and marketing.

Table 1. Students' Record Its Attributes, Description and Possible Variables.

Variables	Descriptions	Possible Values
Gender	Students gender	{Male, Female}
Last School Attended	The originating school of the students	{Home-grown, Transferred-in from Private, Transferred-in from Public}
Subjects	The subjects taken by the students	{Filipino, Araling Panlipunan, MAPEH, FCL, English, Science, Math, TLE, Financial Literacy, Personal Development, Computer}
POcc	Parents' Occupation	{Business, Employed}
Grade	Rating of the students on different subjects	{70%-100%}
GenAv	General Average of the students	{70%-100%}
APer	Final Academic Performance of the students	{90% above- Outstanding, 85%-89%-Very Satisfactory, 80%-84%-Satisfactory, 75%-79%-Fair Satisfactory, 74%-below-Beginning}
Marketing	Options of how students knew about perpetual	School campaign, Parent/s is/are University employee, Parent/s alumni, Through a friend, Through a relative

4. Results and Discussions

The data extracted from the database of the school system was cleaned, modified and mined based on the objectives presented. The model used to analyze the data is RapidMiner employing the data mining technique which is classification using Naïve Bayes.

Figure 1 shows the general average of the students when grouped according to gender. The chart displays that there were lot of female students who had an average between 86%-87% in Grade 7. Most females have an average grade of 80%-93% while male mostly got an average of 76%-89%. However, a male student outnumbered those females who got 95% as general average.

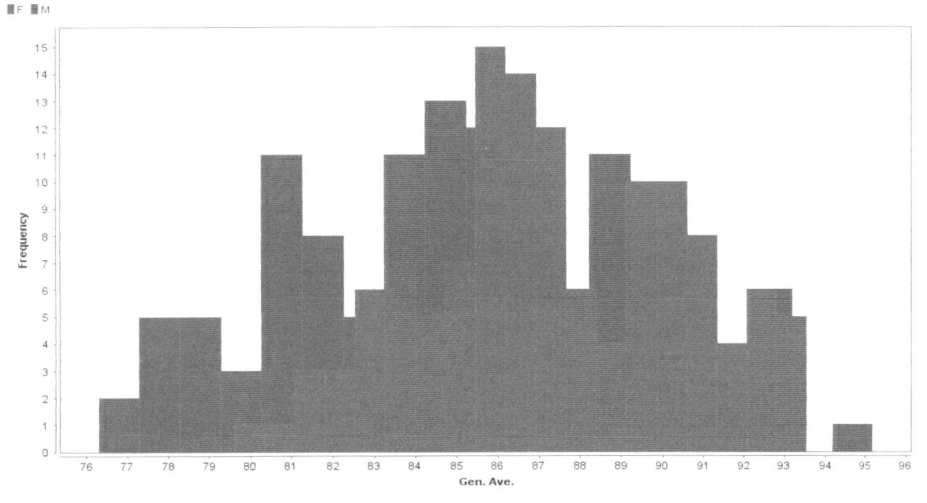

Figure 1. General Average of the Students When Grouped According to Gender.

Figure 2 shows the academic performance of the Grade 7 students based on their gender. The chart explained that mostly of the female students were very satisfactory in terms of academic performance compared to those male students, these were the students who got a general average of 85%-89%. Outstanding students were mostly female also and those who were satisfactory and fairly satisfactory were boys.

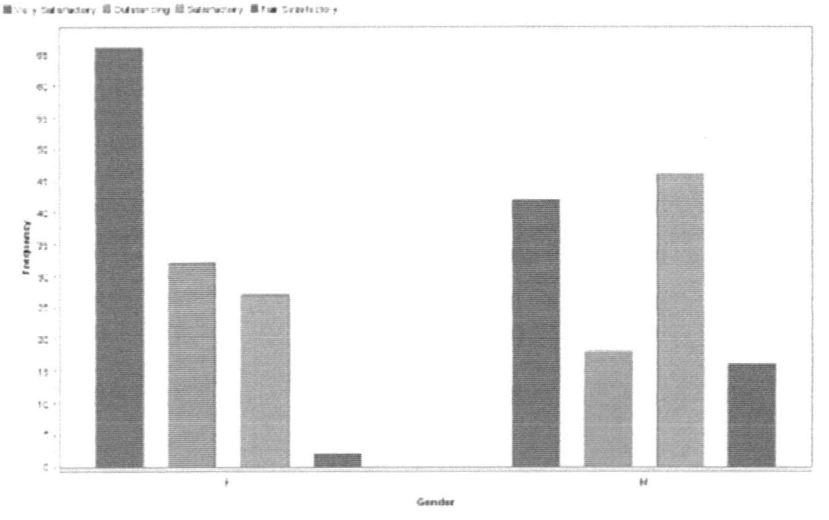

Figure 2. Academic Performance of Male and Female.

Figure 3 shows the distribution of the grades per subject. The chart indicates the subject that most of the students excel most is Personal Development. Based on the general average of the students, Personal Development subject has the highest percentage of passing compared to other subjects while students had difficulty in Araling Panlipunan, most of them got low grades in the said subject.

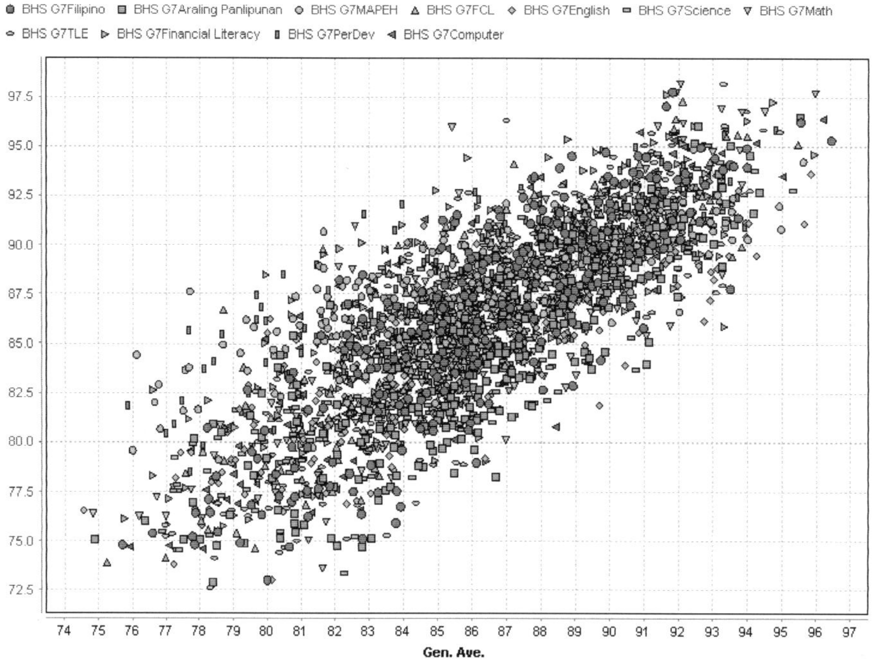

Figure 3. Grades of Students Per Subject.

Figure 4 shows the general average of students when grouped according to last school attended. The chart displays that the students who transferred-in from private schools had highest average obtained in Grade 7 compared to those who were homegrown and transferred-in from public.

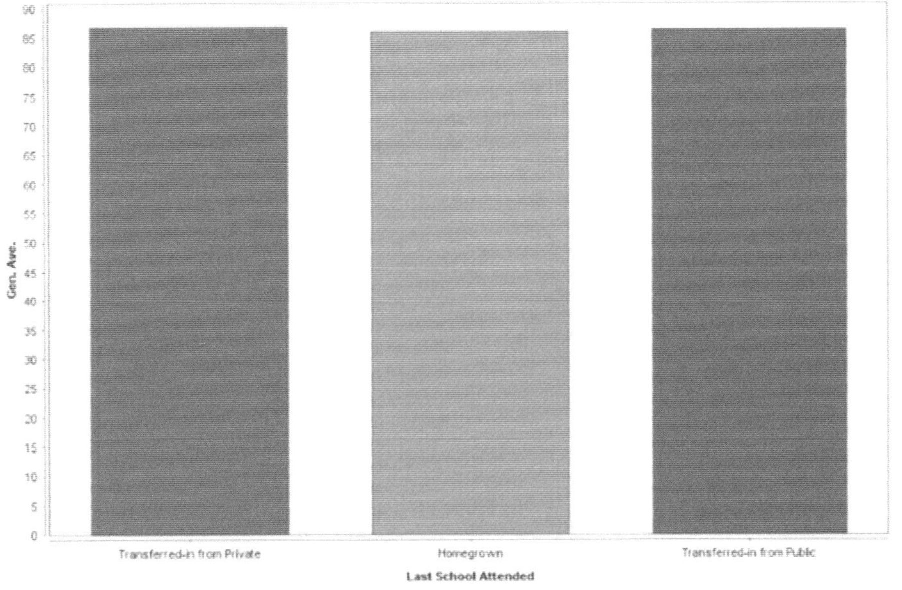

Figure 4. General Average When Grouped According to Last School Attended.

As indicated in figure 5, students whose parents were businessmen had obtained higher general average in Grade 7 compared to those students whose parents were employees. Businessmen parents had time freedom compared to those parents who are employees. They had all the resources that they can provide to their children; time and money.

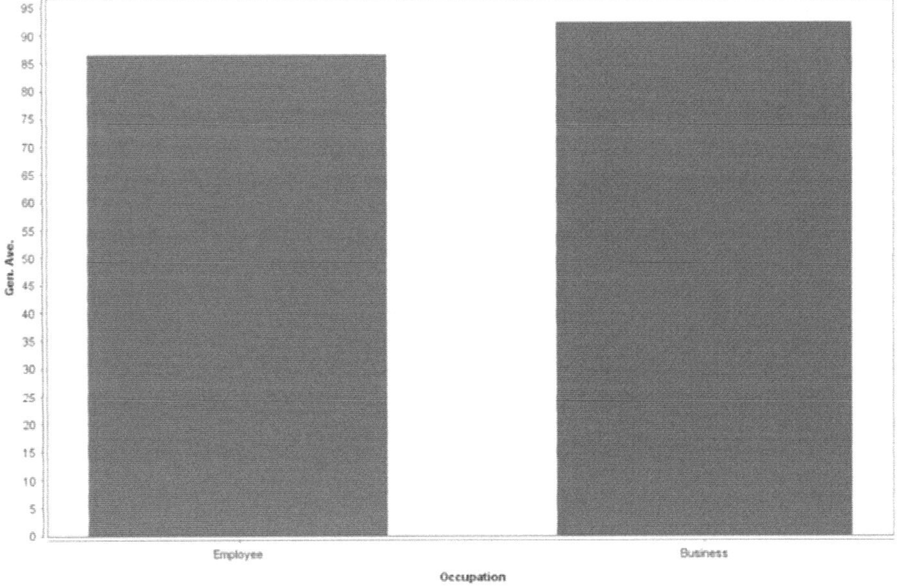

Figure 5. General Average When Grouped According to Parents' Occupation.

The figure shows the predictive analysis of the students' data in terms of their last school attended. The data shows that the accuracy of the model is 35.73% based on the last school attended. Class precision for Transferred-in from private school is 40.74%, 40.21% from the Homegrown school and a class precision of 13.64% is the Transferred-in from the Public school. The results show that there is a high possibility of analysis that students who will enroll will come from the private schools.

Table View Plot View

accuracy: 35.73% +/- 7.42% (micro average: 35.74%)

	true Transferred-in from Priv...	true Homegrown	true Transferred-in from Pu...	class precision
pred. Transferred-in from Pr...	44	39	25	40.74%
pred. Homegrown	36	39	22	40.21%
pred. Transferred-in from P...	15	23	6	13.64%
class recall	46.32%	38.61%	11.32%	

Figure 6. Class Precision in Terms of Last School Attended.

The figure shows the class precision and the class recall of data based on the marketing strategy used by the school. The data sets used here were school campaign, parent/s is/are University employee, through a friend, parent/s alumni and through a relative. These were the options given as to what how the students know about Perpetual. Class precision refers to the ratio of correctly predicted positive observations to the total predicted observations which is shown that Parent/s is/are University employee has the highest percentage of precision of 40%, the same thing is with the class recall which has a percentage of 50%. Recall is the ratio of correctly predicted positive observations to the all observations in actual class-yes. Thus the prediction of data shows that mostly of the students enrolled in Perpetual is because their parents are employees.

accuracy: 30.97% +/- 7.82% (micro average: 30.92%)

	true School Camp...	true Parent/s is/ar...	true Through a frie...	true Parent/s alumni	true Through a rel...	class precision
pred. School Cam...	8	6	5	1	6	30.77%
pred. Parent/s is/a...	5	30	18	7	15	40.00%
pred. Through a fri...	16	13	25	9	14	32.47%
pred. Parent/s alu...	3	4	5	0	6	0.00%
pred. Through a re...	10	7	18	4	14	26.42%
class recall	19.05%	50.00%	35.21%	0.00%	25.45%	

Figure 7. Predictive Analysis of Marketing Strategy.

5. Conclusions

In this study, a model was developed based on some selected input variables. Data mining classification algorithm Naïve Bayes was applied to predict the academic performance of students based on the previous years' database. The tool Rapid miner is used for exploration, statistical analysis and mining of student data. Cross-Validation operator is used to perform a cross-validation process. From the above analysis, the researchers concluded that the Naïve Bayes produces accuracy 92.37% that shows it is possible to obtain a good prediction model. The proposed methodology can be adopted to predict the performance of students and help the teachers and the management to enhance the quality of learning and student's academic performance by doing important decision at the right time. The predictive analysis based on the data could help the school intensify the marketing strategies to encourage lot of students from the community. In the future, the study can be enhancing by including the data with more information about the students and of higher quality which might help to improve the current model performance and also to obtain more accurate student performance.

References

[1] Arockiam L, Charles S, Arulkumar, et al. 2010 *Int. J. Computer Sci. Eng.* Deriving Association between Urban and Rural Students Programming Skills, **2(3)** 687-90.

[2] Baker R S, Corbett A T and Koedinger K R 2004 *7th Int. Conf. Intell. Tutoring Syst.* Detecting Student Misuse of Intelligent Tutoring Systems, **3220** 531-40.

[3] Baker R S, Corbett A T and Koedinger K R 2004 Detecting Student Misuse of Intelligent Tutoring Systems. Proceedings of the International Conference on Intell. Tutoring Syst. 3220 531-40.

[4] Romero C and Ventura S 2010 *IEEE Trans. Syst. Man Cyber. C. (Appl. Rev.)* Educational Data Mining: A Review of the State-of-the-Art, **40(6)** 601-18.

[5] AZIZ A. A., ISMAIL N.H., AHMAD F. and HASAN H 2015 A Framework for Students' Academic Performance Analysis using Naïve Bayes Classifier in Proceedings of the ICIDM.

[6] Priya K S and Senthil Kumar A V 2013 *Int. J. Adv. Netw. Appl.* Improving the Student's Performance Using Educational Data Mining, **4(4)** 1680-5.

[7] Goyal M and Vohra R 2012 *Int. J. Comput. Sci. Issues* Applications of Data Mining in Higher Education, **9(2)**.

[8] Cortez P and Silva A 2008 *EUROSIS-ETI* Using Data Mining to Predict Secondary School Student Performance.

[9] Ramaswami M, Bhaskaran R and CHAID A 2010 *Int. J. Comp. Sci. Issues* Based Performance Prediction Model in Educational Data Mining, **7(1)**.

[10] Dimitoglou G, Adams J A and Jim C M 2012 *J. Comp.* Comparison of the C4.5 and a Naïve Bayes Classifier for the Prediction of Lung Cancer Survivability, **4(8)**.

The Optimization the Latching Mechanism Design Based on Progressive Damage Analysis

YANG Yang[1], LIU Ning[1], SUN Qilong[1], WANG Xiaoxue[1] and CHEN Juwei[1]

[1]Beijing Institute of Space Long March Vehicle, Beijing, China 100076

sunnybuaa@foxmail.com

Abstract. The rocket latching force is an important parameter, prevent the rocket from falling off and affect the launch of the rocket. In order to obtain the optimal engineering design of the latching bar, this paper predicts the damage failure of two kinds of latching bar, based on the continuous damage theory and the Johnson-Cook constitutive model. Compared with the experimental data, the relative error accuracy of the ultimate strength simulation for rocket latching force is reached 5%.

1. Introduction

The latching mechanism is used to prevent the rocket falling off and ensures that the relative position of the rocket in canister does not change during the transportation and preparation phases. After the rocket is fired and the thrust is generated, the rocket explodes against the latching force. The practice verification shows that the rockets deviation of the impact point angle caused by the deviation of the off-track velocity with uniformly latching force can be reduced [1-3]. The basial latching mechanism used in the rocket launcher has four kinds: friction-latching mechanism, spring-latching mechanism, lever-latching mechanism and shear pin latching mechanism [4-6].

This paper focus on the shear pin latching mechanism, wherein the key component for the latching force is the latching bar. The latching force F_L provided by the latching bar needs to meet the following conditions.

$$\text{Static / Quasi-static:}\quad F_L > n_s F_s; \tag{1}$$

$$\text{Dynamic:}\quad F'_L < n_d F_d. \tag{2}$$

where F_s is the maximum load in the transportation and other ground conditions, n_s is the safety factor corresponding to F_s, F_d is the motor thrust in launch conditions and n_d is the corresponding safety factor corresponding to F_d.

The equation (1) denotes that the latch bar can be adapted to the ground condition, the equation (2) denotes that the latch force does not affect the launch. In order to ensure the uniform latching force, it is necessary to analyze the effect of structure parameters and determinate structural design of the latching bar.

In order to obtain the optimal engineering design of the latching bar, this paper predicts the damage failure of two kinds of latching bar. The progressive damage theory in static / quasi-static is applied to predict the damage load of the 7A04 aluminum alloy latching bar, and simulation data is consistent with the experimental results. According to experimental data, the motor thrust is enough to over the dynamic damage load of the latching bar, and the stresses and strains of the other part in latching m Mechanism should be analyses either.

2. Static / Quasi-static Constitutive Model

Considering the response of a typical metal specimen during a simple tensile test, the stress-strain response, such as that illustrated in figure 1. The material response is initially linear elastic, *a*→*b*, followed by plastic yielding with strain hardening, *b*→*c*. Beyond point *c* there is a marked reduction of load-carrying capacity until rupture, *c*→*d*. The deformation during this last phase is localized in a neck region of the specimen. Point *c* identifies the material state at the onset of damage, which is referred to as the damage initiation criterion. Beyond this point, the stress-strain response *c*→*d* is governed by the evolution of the degradation of the stiffness in the region of strain localization. In the context of damage mechanics *c*→*d* can be viewed as the degraded response of the curve *c*→*d'* that the material has followed in the absence of damage.

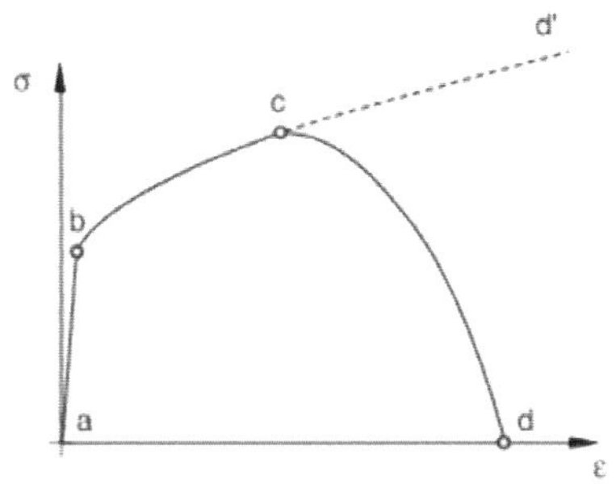

Figure 1. Typical uniaxial stress-strain response of a metal specimen.

the specification of a failure mechanism consists of four distinct parts:

- the Elasticity law (*a*→*b* in figure 1),

$$\varepsilon_e = \frac{\sigma}{E(1-D)} \tag{3}$$

where *E* is Young's modulus of the undamaged material; *v* is Poisson's ratio; *D* received the status of an internal state variable: 0≦*D*≦1 (0 for the undamaged state and 1for failure).

- the Plasticity law (*b*→*c* in figure 1),

In order to model plasticity two kinds of strain hardening are usually considered: (a) the isotropic hardening related to the density of dislocations or flow arrests; (b) the kinematic hardening related to the state of internal microstress concentrations, the corresponding back stress defines the center of the elastic domain in tension compression (or in three dimensions).

This model can be described as

$$\left|\frac{\sigma}{1-D} - X\right| - R - \sigma_y = 0 \tag{4}$$

- a damage initiation criterion (*c* in figure 1),

The microcracks are initiated by the accumulation of microstresses accompanying incompatibilities of microstrains or the accumulation of dislocations in metals. The damage threshold can be defined as

IOP Conf. Series: Materials Science and Engineering **649** (2019) 012019 doi:10.1088/1757-899X/649/1/012019

$$\varepsilon_p < \varepsilon_{pD} \rightarrow D = 0 \tag{5}$$

where ε_{pD} is the damage initiation criterion strain.

- a damage evolution law (e.g., $c \rightarrow d$ in figure 1).

3. Dynamic Constitutive Model

The Johnson-Cook criterion (available only in Abaqus/Explicit) is a special case of the ductile criterion in which the equivalent plastic strain at the onset of damage, $\bar{\varepsilon}_D^{pl}$ is assumed to be of the form

$$\bar{\varepsilon}_D^{pl} = [d_1 + d_2 \exp(-d_3\eta)]\left[1 + d_4 \ln\left(\frac{\dot{\varepsilon}^{pl}}{\dot{\varepsilon}_0}\right)\right](1 + d_5\hat{\theta}) \tag{6}$$

where d_1-d_5 are failure parameters and $\dot{\varepsilon}_0$ is the reference strain rate. This expression differs from the original formula published by Johnson and Cook in the sign of the parameter d_3. This difference is motivated by the fact that most materials experience a decrease in $\bar{\varepsilon}_D^{pl}$ with increasing stress triaxiality; therefore, d_3 in the above expression will usually take positive values. $\hat{\theta}$ is the nondimensional temperature defined as

$$\hat{\theta} \equiv \begin{cases} 0 & \text{for} \quad \theta < \theta_{\text{transition}} \\ (\theta - \theta_{\text{transition}})/(\theta_{\text{melt}} - \theta_{\text{transition}}) & \text{for} \quad \theta_{\text{transition}} \le \theta \le \theta_{\text{melt}} \\ 1 & \text{for} \quad \theta > \theta_{\text{melt}} \end{cases} \tag{7}$$

where θ is the current temperature, θ_{melt} is the melting temperature, and $\theta_{\text{transition}}$ is the transition temperature defined as the one at or below which there is no temperature dependence on the expression of the damage strain $\bar{\varepsilon}_D^{pl}$. The material parameters must be measured at or below the transition temperature.

4. Experimental Design

The carrying forces of the latching bars with two different characteristic parameters were examined in the test to obtained the latching force, and the test diagram was shown in figure 2. By the loading bulk, the loading was concentrated on the middle part of the latching bar, and the breaking force value, fracture displacement and fracture morphology of the latching bar were recorded. Two kinds of bars are tested and the structural and material characteristics of the latching bars were shown in figure 3 and table 1. The materials of all the part in the test system were steel, except the latching bar.

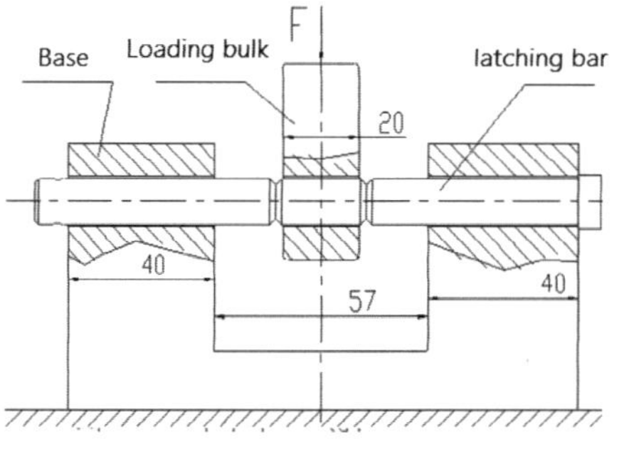

Figure 2. The carrying capacity test of the latching bar.

IOP Conf. Series: Materials Science and Engineering **649** (2019) 012019 doi:10.1088/1757-899X/649/1/012019

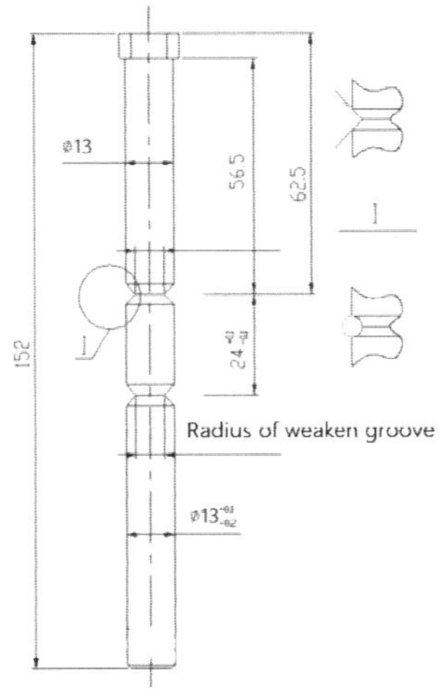

Figure 3. The structural characteristics of the latching bar.

Table 1. The structure parameters of latching bar.

Number of scheme	Quantity	Radius of weakening groove/mm	Type of weakening groove	Material
Scheme 1	3	7.5	90°	7A04 aluminum alloy
Scheme 2	3	7.5	R=2mm	7A04 aluminum alloy

5. Numerical Simulation

The numerical model was established as shown in figure 4, including three parts, i.e. the base, loading bulk and latching bar. In the model, the bottom of the base was fixed as the boundary condition and only loading direction freedom of loading bulk was free. The real contact relationships between the latching bar, the base and the loading bulk were defined. The simplified loading rod model was used to simulate the test loading device. The latching bar was defined as the ductile metals, and the materials properties was according to the reference [5], wherein the stress triaxiality and the damage initiation criterion has been defined. Assume other parts in this test was linear elasticity materials.

Figure 4. The numerical model of test.

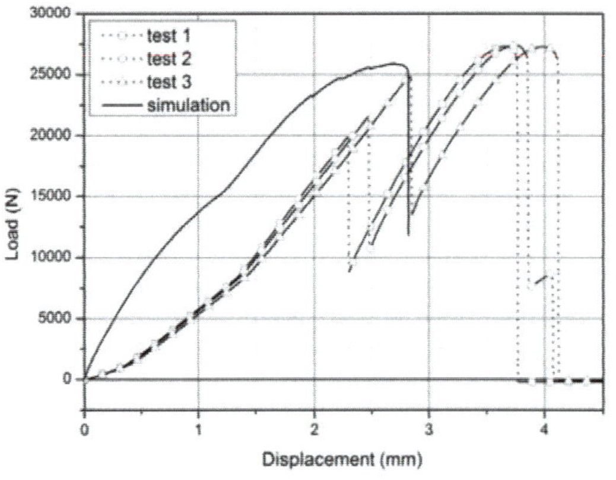

Figure 5. The load-displacement curve of latching bar with V weaken groove.

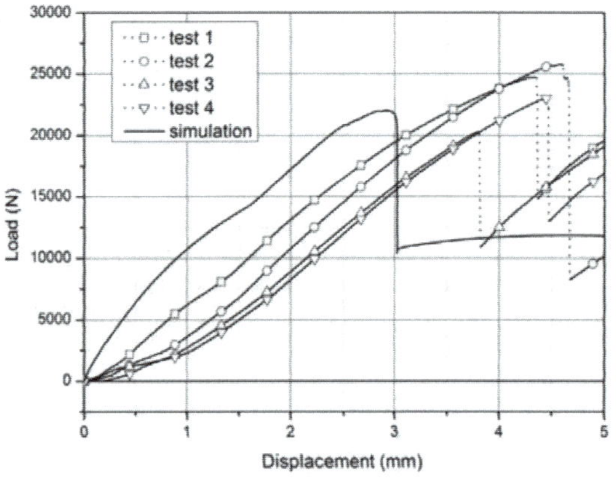

Figure 6. The load-displacement curve of latching bar with circle weaken groove.

The result of carrying capacity of the latching mechanism was given, as shown in figure 5 and figure 6. The failure mode of different latching bars was yield failure mainly. The latching force was related to type of weakening groove and the dimension of the latching bar directly.

By comparison, it was considered that the latching force of the two schemes were both satisfied the *condition* (1), and meanwhile the damage phenomenon was different for the two type of weakening groove. Though all the load-displacement curve had two peaks, the maxmum peak load with Scheme 1 was the second one and the maxmum peak load with Scheme 2 was the first one. It was implied that the displacment related to the peak load was smaller in the Scheme 2. Thus, the latching force of Scheme 2 it was more suitable for the environmental requirements because of its uniform and better properties. Therefore, Scheme 2 was selected.

The prediction of failure load of latching bar was consistent with the experimental results, but the prediction of displacement was affected by the systematic error of the experimental machine.

In order to check the adaptation of latching bar for condition (2), the dynamic behavior of whole mechanism was simulated by applying the "J-C Constitutive". The FEM model established for the local structure as shown in figure 7, wherein the red flat-plate and white bulk were represented part of the launch canister body and rocket respectively.

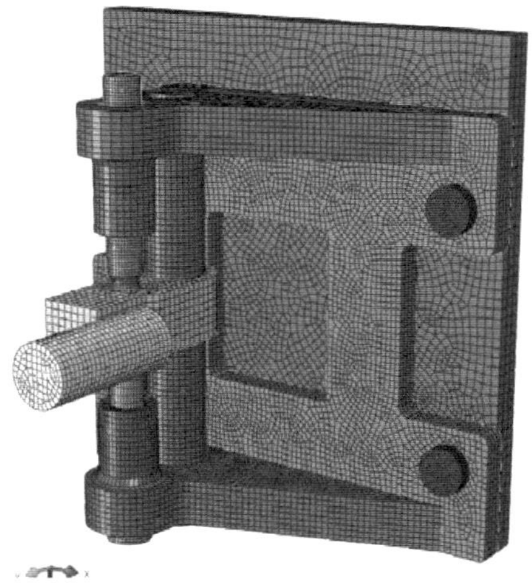

Figure 7. The numerical model of latch mechanism (local).

According to the calculation results as shown in figure 8, the maximum of stress is 361.8 MPa, which was small enough to meet the condition (2).

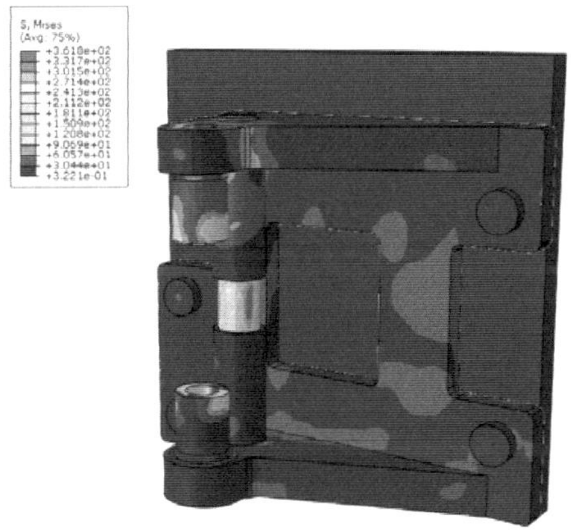

Figure 8. The stress of the latch mechanism (local).

By the practice verification, the new latching mechanism can adapt to the environment of rocket explodes and the simulation was correct.

6. Conclusion

In this paper, the damage load of the rocket latching bar was simulated based on the strength failure model, and the conclusion is obtained as followed.

1) The failure mode of latching bar was yield failure mainly. The equivalent plastic strain is the key parameter to judge the damage initiation criterion. When the equivalent plastic strain of the structural member reaches the material limit, the structural damage is occurred.

2)The displacment related to the peak load was smaller in the Scheme 2. Thus, the latching force of Scheme 2 it was more suitable for the environmental requirements because of its uniform and better properties.

3) In this paper, the prediction of failure load of latching bar based on the nonlinear damage constitutive was consistent with the experimental results, but the prediction of displacement is affected by the systematic error of the experimental machine.

4) By the practice verification, the new latching mechanism can adapt to the environment of rocket explodes and the simulation was correct.

References

[1] LIN Y M and RAO J 1997 *J. Ballistics* Model establishment and numerical simulation investigation on launch dynamics of rocket flexible systems, **4** 38-42.

[2] WANG Z J and ZHAO W X 1996 *J. Ballistics* Numerical simulation on the initial disturbance of rocket, **3** 74-9.

[3] DU C B and HUANG L 2006 *J. Nanjing University Sci. Technol.* Theoretical study and test on thrust misalignment of solid rocket motor, 550-3.

[4] LIU W and SONG X H 2017 *J. Projectiles, Rockets, Missiles Guidance* Numerical Analysis and Experimental Verification for Locking Force of Rocket Artillery, **5** 151-5.

[5] ZHANG W, XIAO X and WEI G 2011 *EXPLOS. SHOCK WAVES* Constitutive relation and fracture model of 7A04 aluminum alloy, **1** 81-7.

[6] WANG J, WANG C H, YANG Y and SUN X C 2016 *STRUCT. ENVIRON. ENG.* Progressive damage analysis of C/SiC composite shaft sleeve, **43(5)** 30-7.

Study on Separation Dynamics of Explosive Bolt and Test Verification

XIN Kehao[1], LIU Sai[1], ZHU Xiaojun[1], CHEN Jingyi[1] and GU Xiaolong[1]

[1]Beijing Institute of Space Long March Vehicle, Beijing, China 100076

xinkh_bise@163.com

Abstract. In order to simulate the dynamic process after the separation of explosive bolt, and to analyze whether the impact force of the explosion could cause the damage of the explosive bolt box structure. The explicit dynamic method of ABAQUS was adopted in this paper to carry out simulation analysis on the process of the explosive bolt passing through the corner piece and hitting the buffer block after unlocking, and to evaluate the strength and deformation of the bolt box under impact load. The verification test of explosive bolt was carried out to evaluate the reliability of the design, and the test results indicated that the designed explosive bolt box achieves a successful acceptance of the explosive bolt of the separation with design margin, the structure of the box does not have obvious deformation under the impact load, and the strength design of the explosive bolt box meets the use requirements.

1. Introduction

The explosive bolt, cutting pyrotechnic and release actuator are used in aerospace models for separation, and the explosive bolt is the most widely used due to its stable connection and high reliability [1, 2]. Wu simulated the process of PETN exploding and its effect to the explosive bolt box by nonlinear dynamic finite element analysis software LS-DYNA [3]. Yang used explicit dynamic simulation method to analyze the separation process by Dytran software, and the shock response spectrum (SRS) at different measurement points was investigated [4]. The total process of the explosive bolt passing through the flake spring and striking the cushion block was simulated dynamically with ABAQUS by Wang [5].

During the process of separating and unlocking the explosion bolt, extraneous materials are generated due to the action of the pyrotechnics, and the explosion bolt rod would obtain a higher separation speed. If the structural strength design of the explosive bolt box body is insufficient, or the buffer block cannot absorb the separation energy, the structure of the box may be broken after being impacted by the bolt rod, which may cause the separation of products to enter the aircraft, and seriously lead to the failure of the flight test. Therefore, in the design of explosive bolt box, not only the strength of the box body is sufficient, and the structure would not be damaged due to pyrotechnic impact during the separation process, the bolt rod and the separated products should be reliably stored. In the past design process, most of the mature design experience was used for reference, and combined with the multi-round tests for verification, the analysis of the separation process of explosive bolt box is relatively lack, which affects the design and development process.

In this paper, the dynamics simulation analysis method was used to simulate the separation process of the explosive bolt. The storage performance of the explosive bolt box is evaluated, and the separation test of the explosive bolt has been validated for the structural design and the simulation results.

2. Design of Explosive Bolt Box Structure

Conventional explosive bolt box is designed with corner piece, buffer block and other structures. Through the cushioning action of the corner piece and the buffer block, the separation energy of the bolt rod caused by the action of the pyrotechnic product can be absorbed, and the bolt box caused by the impact can be avoided. In addition to the above structure, another typical explosive bolt box structure is provided with flake spring on both sides of the box body, the flake spring could absorb a part of the impact energy in the process of bolt rod separation, and can also play a role in the bolt rod rebound clamping, avoid bolt in the separation of protruding separation surface. Typical explosive bolt box structures are shown in figure 1.

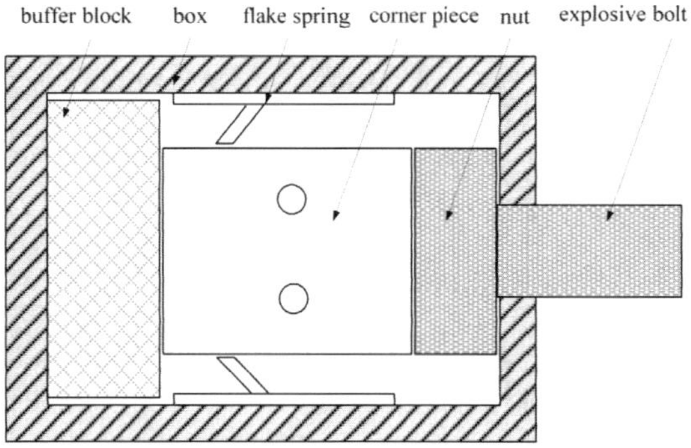

Figure 1. Schematic of the explosive bolt box structure.

The working process of the explosive bolt box is briefly described as follows: the explosive bolt rod is cut off under the action of pyrotechnical products and generates a certain separation speed, the nut moves in the box. By properly designing the structure of the corner piece and the buffer block, it can fully absorb the separation energy of the bolt rod assembly, and finally stop moving in the box, so as to realize the storage of bolt rod and nut.

3. Dynamics Simulation Analysis

Explosive bolt involves a lot of structures in the separation process, and the motion process is complicated. It is difficult to accurately simulate the entire separation process and structural strength by means of engineering analysis. Therefore, the ABAQUS was used to simulate the whole separation process by dynamics simulation analysis method in this paper, the strength of the box body and components was evaluated, whether the box body structure met the design requirements at the maximum separation speed was analyzed.

3.1. Explicit Dynamics Simulation Analysis Method

The dynamic equation an object under the action of external force F can be expressed as:

$$Ma + Cv + Ku = F \tag{1}$$

In equation (1), M. C, K are the mass, damping and stiffness matrices of the object.
The speed of the object at time t_n is:

$$a_{(n)} = M^{-1} (F_{(n)} - Ku_{(n)} - Cv_{(n)}) \tag{2}$$

For the above equations, the solution to the dynamics is to use the central difference method, which assumes that the acceleration of the motion of the object is constant over a time step, i.e.

$$v_{(n+1/2)} = v_{(n-1/2)} + a_{(n)} (\Delta t_{(n+1/2)} + \Delta t_{(n-1/2)}) / 2 \tag{3}$$

$$u_{(n+1)} = u_{(n)} + v_{(n+1/2)} \Delta t_{(n+1/2)} \tag{4}$$

Substituting equation (3) and equation (4) into equation (2), the acceleration at time t_n could be solved.

3.2. Impact Speed Analysis

When the explosive bolt is normally separated, the impulse I generated by the pyrotechnics is 5.25 Ns. According to the momentum theorem, the initial speed of the bolt and the nut before the separation is 0. After the separation, the bolt and nut are moved to the inside of the box at the same speed.

$$I = mv \tag{5}$$

In equation (5), I is the impulse that generated by the separation and unlocking of pyrotechnics, m is the mass of bolt and nut, a total of 89g, and v is the initial speed after the separation of bolt and nut.

It can be calculated that the initial speed of the bolt and nut is 59 m/s (100% speed) under the designed impulse of 5.25 Ns. The separation speed is 71 m/s (120% speed) considering a certain design margin.

3.3. Analysis of Action of Flake Spring

The function of the flake spring is to block the nut when the bolt rod and the nut move toward the inside of the box during the separation process, absorb a part of the separation energy, slow down the movement speed of the rod and the nut. After the buffer block has absorbed the energy, the explosive nut does not move in the opposite direction to protrude the separation surface by the angle that the flake spring is opened. The flake spring is spring steel, and the mechanical properties at 20℃ are shown in the table 1.

Table 1. Material property of flake spring.

E/GPa	μ	σ_b/MPa
206	0.29	1570

The initial speed calculated in last section is used as the separation speed of the bolt rod. The separation process of the two structural design forms with or withnot flake spring is simulated by ABAQUS software. The movement speed of the bolt rod and the nut was extracted from the simulation analysis results. It can be seen from the curve of the movement speed of the bolt and the nut shown in figure 2 that there is no significant differences in the movement speed of the bolt rod and the nut under the two structual forms. It is indicated that the flake spring does not fully absorb the energy of the bolt during the movement of the bolt and does not affect the bolt rod movement track.

At the same time, the movement process of the bolt and the nut is analyzed after the buffer block absorb the separation energy. In both cases, the energy of the bolt rod and the nut is absorbed by the buffer block and the corner piece. The rebound speed of the bolt rod and nut decreased from 6.5 m/s to 5.9 m/s, and neither of the two structual forms the bolt and nut would not protrude from the separation surface in the end.

IOP Conf. Series: Materials Science and Engineering **649** (2019) 012020 doi:10.1088/1757-899X/649/1/012020

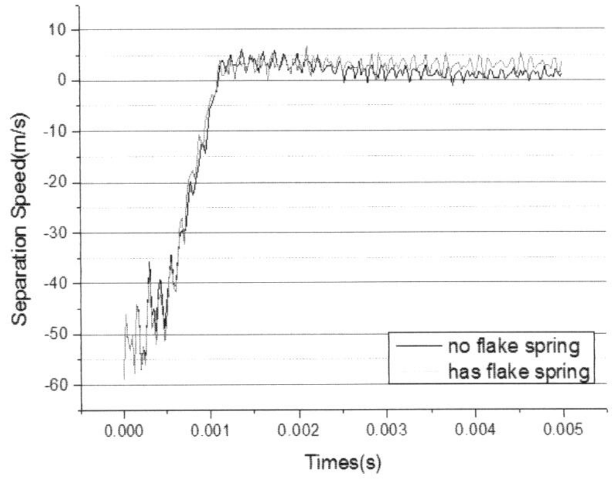

Figure 2. Movement speed of the bolt rod and the nut under the two structual forms.

3.4. Stress Analysis of Explosive Bolt Box

The explosive bolt would generate impact load during the separation process. During the use process, the damage of the box structure must be avoided. Therefore, a buffer block is installed inside the box, and the buffer block should effectively absorb the impact energy by bolt, nut and corner piece.

The finite element simulation software was used to simulate and analyze the stress situation of the explosive bolt box under the condition of 100% and 120% speed. The explosive bolt box material is aluminum alloy. The mechanical properties are shown in table 2. The stress of the box is shown in the figure 3.

Table 2. Material properties of explosive bolt box.

E/GPa	μ	σ_b/MPa	σ_s/MPa
70	0.3	280	220

According to the calculation results, the most severe stress area of the box is the contact position between the bottom and the buffer block. At the end of impact separation, the equivalent stress at this position is 221 MPa at 100% speed and 255 MPa at 120% speed, none of situations exceeded the strength limit of the box material. The strength of the box of the explosive bolt box is designed to meet the requirement of use, and has a certain of design margin.

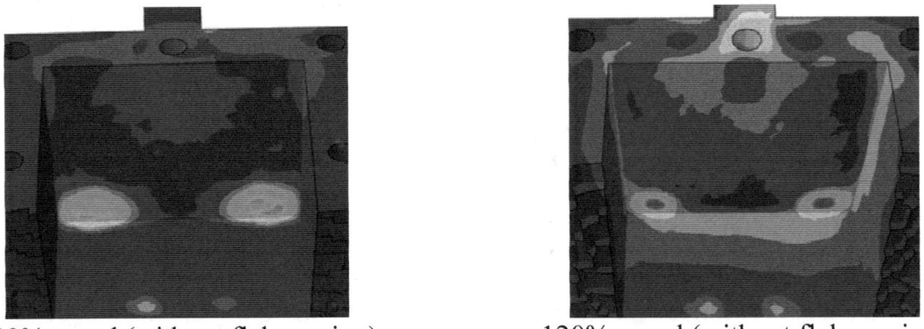

100% speed (without flake spring) 120% speed (without flake spring)

Figure 3. The stress of explosive bolt box.

3.5. Deformation Analysis of Explosive Bolt Box Buffer Structure

From the simulation calculation results in figure 4, the buffer block inside the box is the main part to absorb the bolt and nut movement energy. In order to ensure that the box does not damage due to excessive impact, the buffer block needs to be able to absorb all the energy and leave a certain amount of compression. By simulating the deformation of the buffer block at 100% speed with flake spring, 100% speed without flake spring and 120% speed without flake spring, the buffer block still has a certain compression margin at the maximum design speed, and there is no crushing phenomenon. It is shown that the design form and thickness of buffer block structure meet the design requirements.

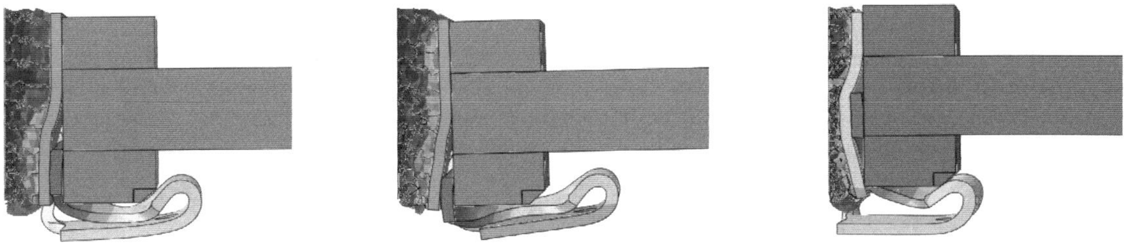

100% speed (with flake spring) 100% speed (without flake spring) 120% speed (without flake spring)

Figure 4. Deformation of explosive bolt box buffer in different structural forms.

3.6. The Verification Test

According to the above simulation analysis, the design of the box structure and buffer block can absorb the separation energy, and the bolt rod does not protrude the separation surface after the separation is completed. Although the flake spring can absorb certain of energy, the box structure is too small, so the actual installation of the flake spring would have difficulties. Considered the actual product design, the flake spring was not to install finally.

The verification test of the explosive bolt box was carried out according to the above structure. The air gun was used to load the impact, and the compressed air was used to make the projectile produce the set speed. When the projectile hits the bolt rod, the separation speed generated by the explosion of the pyrotechnic product is simulated.

100% speed 120% speed

Figure 5. Results of verification test with different speed.

The test simulated the separation, impact and storage of the explosive bolt box, and tests were carried out in two working conditions of 100% speed and 120% speed. Under all tests the bolt rod did not protrude from the separation surface, and no obvious deformation was found in the box structure, indicating the reliability of the box structure design. After the test, the compression of the buffer block is basically consistent with the simulation results, and a certain amount of compression is still maintained.

4. Conclusion

IOP Conf. Series: Materials Science and Engineering **649** (2019) 012020 doi:10.1088/1757-899X/649/1/012020

In this paper, explicit dynamic analysis method is used to simulate the whole process of the impact of the corner piece on the buffer block after the separation of the explosive bolt, and the strength of the explosive bolt box under the impact load is evaluated. According to the separation speed designed air gun impact test, test results verified the effectiveness of the explosive bolt box structure strength design and simulation results.

a) According to the simulation and test results, the corner piece and the buffer block in the explosive bolt box are the main parts to absorb the explosion separation energy. The spring flake has little effect on absorbing the explosion separation energy.

b) At 100% speed, the maximum stress on the explosive bolt box is 221 MPa, which is not larger than the strength limit of the box material, and there was no obvious gap between the bolt box and cover plate. At 120% speed, there is no obvious deformation of the box and the cover plate of the bolt, which verifies the structural design of the bolt box.

c) The high-speed impact generated by the air gun can better simulate the separation speed generated by the explosion of the separation bolt. Compared with the deformation of the buffer block in the bolt box after the actual explosion separation and air gun test, the amount of deformation is basically the same. It is indicated that the air gun test can use to replace the actual explosive bolt separation test, and it can reduce test expenses and develpment shchle.

d) Using ABAQUS explicit dynamics method, the whole process of explosive bolt separation can be simulated realistically, and the simulation results are consistent with the test results. The process of the simulation simplifies the design process, and saves part of the test costs, which can provide a reference for subsequent design of explosive separation works.

References

[1] Chunquan H E, Zuo Y and Yaokun Y E 2009 *Spacecraft Recovery Remote Sens.* Study of Stage Separation Device for Missile, **(3)** 70-7.

[2] Holden M S, Harvey J, MacLean M and Wadhams T 2005 *AIAA* Development and Application of a New Ground Test Capability to Conduct Full-Scale Shroud and Stage Separation Studies at Duplicated Flight Conditions, 694-6.

[3] Wu Y H, Wang X H, Ma B J, et al. 2007 *Struct. Environ. Eng.* The simulation of explosion and the calculation of the impact strength of the explosive bolt box, **34** 10-5.

[4] Yang N, Fan B J and Hong Q F 2017 *Manned Spaceflight* Numerical Analysis of Shock Response in Typical Low Impact Separation Nut, **23** 776-83.

[5] Wang Z W, Wu H W, Yang K, et al. 2018 *Missiles Space Veh.* Research on the Dynamics Simulation of the Explosive Bolt Separation, **359** 111-21.

2019 World Symposium on Smart Materials and Applications (WSSMA 2019) IOP Publishing
IOP Conf. Series: Materials Science and Engineering **649** (2019) 012021 doi:10.1088/1757-899X/649/1/012021

The Method to Measure the Body's Local Temperature: What Materials Should Be Used to Fix The iButton on the Skin

Tianyu Xi[1], Huan Qin[1], Xiaowei Lv[1] and Qiaochu Wang[1]

[1]School of Architecture, Harbin Institute of Technology, Harbin P.R. China

tonny_kid@126.com, qinhuan0401@foxmail.com, 656727544@qq.com, 1364491243@qq.com

Abstract. iButton, a wireless temperature system, is currently the most popular and inexpensive miniature temperature datalogger commercially available [1]. However, there is no any previous study to evaluate the effect of iButtons' fixed material on the accuracy of thermal measure. This article carried an experiment to test the impacts of four different fixed materials on iButtons: transparent adhesive tape, Aluminum foil paper, black and white bandages. The result shows that transparent adhesive tape can reduce the influence of wind speed while Aluminum foil paper can prevent solar radiation effectively. As for bandages, the temperature of the black bandages group is higher than the white bandages group which indicated that black material can absorb more solar radiation. However, the function of bandages to reduce the impact of ambient heat is not obvious in this test, though it is existed.

1. Introduction

Outdoor thermal comfort is a factor of vital importance in environmental comfort evaluation. Over the last 2 decades, many scholars have been dedicated to investigating outdoor thermal comfort. And in these studies, besides the micrometeorological factors, people's skin temperature has gradually become an important part in outdoor thermal comfort researches. As the barrier of human body, skin plays an important role in heat transfer between the human body and the thermal environment. And in order to quantify the thermal transmission between human body and environment, it's essential to measure the human skin temperature. The temperature distribution over the body's surface provides useful information for many research and clinical applications [2].

For the aim of skin temperature measuring, various kinds of equipment have been used by other researchers. For example, Hui Zhang et al. (2010) used an ingestible radio-thermometer pill to measure the body local temperature [3]. Andrei Claudiu Cosma and Rahul Simha [4] obtained the data by their sensor fusion which was consisted by a depth sensor and color camera combination (Kinect 2), a thermographic camera (Flir Lepton) and a point IR sensor (MLX90614). R. Califano et al. (2017) adopted the temperature measurement system which consisted of fourteen 100 Ohm Platinum RTDs (resistance temperature detectors), four RTD analog inputs, each with four channels (Type 9217, National Instruments Italy S.r.l) [5].

In this article, iButtons which showing to be valid and reliable devices [6] are being introduced for skin temperature measure. In fact, this system has already been used in many studies and participants could wear 4–8 [7] or even 26 or more iButtons attached to the skin [8] during the experiment. For example, Cong Song et al. (2010) attached the iButton (DS1922L Thermochron, Maxim, USA) to skin-

IOP Conf. Series: Materials Science and Engineering **649** (2019) 012021 doi:10.1088/1757-899X/649/1/012021

measuring sites to measure subjects' local skin temperature and stitched the iButton (DS1923 Hygrochron, Maxim, USA) on the surface of bed covers to measure the bed climate [9]. The accuracy and precision of the data recorded by iButtons has been proved by many researchers such as E. Rubinstein et al., 1990[10], Angilletta and Krochmal, 2003[11], Dechmann et al., 2004[12] and Isaac et al., 2008[13]. Though iButton can record precise temperature data, its accuracy may be potentially affected by the materials which used to fix the devices on the skin. However, there is no any study to evaluate the accuracy of iButton when they are attached on the skin by different materials. The objective of this study is trying to compare the material's effect on iButtons when different materials are applied to fix the devices and we hope to find out the best material for the iButtons to get the precise data.

2. Materials and Methods

Our study evaluated whether different materials which are used for fixation can affect the accuracy of iButtons and to what extent the iButtons can be affected. The fixation materials we use in this study are transparent adhesive tape, Aluminum foil paper, black and white bandages respectively. As for the reason why, we choose the four types of material above is that in order to obtain accurate skin temperature data during outdoor experiments we must eliminate the influence of the external environment factors such as wind, solar radiation and ambient heat (air temperature). And the transparent adhesive tape can reduce the impact of wind effectively. Accordingly, the Aluminum foil paper can prevent the solar radiation while the bandage can reduce the influence of air temperature. Before we start the test, we have designed several experimental control groups for comparison, which is shown in table 1.

Table 1. Material and Method to fix the devices AND Their Function Abilities.

Point Number	Materials	Wind Prevention Ability	Radiation Absorbtion Ability	Thermal Isolation Ability
1	black bandage (2 layers)	weak	strong	weak
2	black bandage (4 layers)	weak	strong	normal
3	black bandage (6 layers)	weak	strong	strong
4	white bandage (2 layers)	weak	normal	weak
5	white bandage (4 layers)	weak	normal	normal
6	white bandage (6 layers)	weak	normal	strong
7	white bandage (2 layers) + Aluminum foil paper	strong	weak	weak
8	white bandage (4 layers) + Aluminum foil paper	strong	weak	normal
9	white bandage (4 layers) + Aluminum foil paper	strong	weak	normal
10	white bandage (6 layers) + Aluminum foil paper	strong	weak	strong
11	white bandage (2 layers) + transparent adhesive tape	strong	normal	weak
12	white bandage (4 layers) + transparent adhesive tape	strong	normal	normal
13	white bandage (6 layers) + transparent adhesive tape	strong	normal	strong
14	Medical rubber (2 layers)	normal	normal	normal

The experiments were performed on a five-story room's windowsill which is made of thick granite. And during the experiments, we made sure that there is an ideal testing condition of good quality of ventilation and sunshine. Figure 1 shows different materials and methods we used to fix the instruments.

Figure 1. Various kinds of materials to fix iButtons.

We collected the data on February 17th 2019 in Harbin (126.63E, 45.75N). The weather was sunny, and the air temperature ranges was between -2℃ and 10℃. In order to start the devices at 0 o'clock, we fixed the iButton at February 16th night. Also, in order to make sure that the windowsill surface would be in a stable condition, we opened the windows in advance and kept it opening in the whole day during the experiment.

3. Result and Discussion

Figure 2 shows the changes of temperature in nearly a whole day. Although all of the curves have the similar trend, their values are significantly different from each other. For instance, line-1, line-2, and line-3 have the lowest temperature value in the afternoon and night while during the morning they have the highest temperature. On the contrary, the line-14 is higher than other lines during the period of afternoon and night while during the morning it turns into the lowest one. The rest of the lines are almost indistinguishable from each other in the whole day.

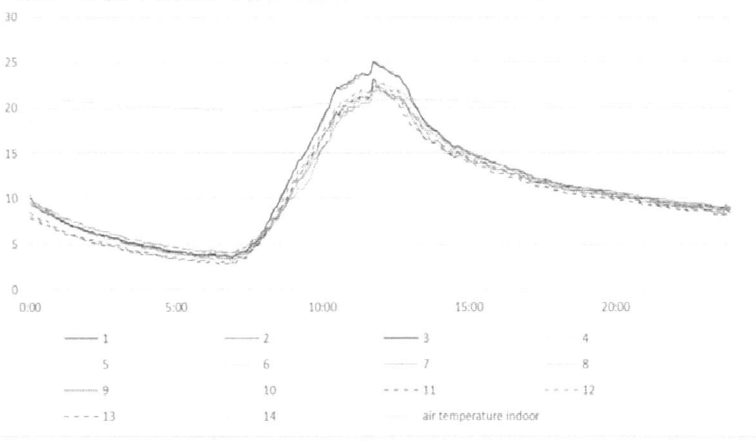

Figure 2. Experiment data from all instruments.

3.1. Result-1 (Bandage Color Analysis)

In order to find the influence of solar radiation on the function of the instrument, we compared the following sets of experimental data. As can be seen from figure 3, the six lines can be approximately divided into two parts. One part (line-1, line-2 and line-3) which uses the black bandages has a higher temperature than the other part (line-4, line-5, line-6) which uses the white bandages, that is because the black material can absorb more solar radiation than the white material.

As is shown in figure 3, there was no significant differences in all lines when there is no sunshine before the sun rose and after 1 o'clock afternoon. During the period between 7:00 a.m. and 13:00 p.m., the sunshine is stronger than other time in the whole day. The thermometers fixed by black bandages reached its peak point at 11:45am, about two degrees Celsius higher than the white bandages group, and this is because of the stronger ability of black materials to absorb the solar radiation.

3.2. Result-2 (Radiant Prevention Analysis)

In figure 4, in order to verify the conclusion of Result-1 and measure what extent of solar radiation can influence the iButton's work, one more group data is added in the figure which used the Aluminum foil paper to prevent the impact of the solar radiation. In the morning, the white bandage group (line-4, line-5, and line-6) is in higher temperature while the Aluminum foil paper group (line-7, line-8 and line-9) is in lower temperature. At 11:45 am, the white bandage group (line-4, line-5 and line-6) reached the highest peak points which are 22.7℃, 23℃ and 22.8℃ respectively while the peak points of Aluminum foil paper group (line-7, line-8 and line-10) at the same time are 21.8℃, 21.6℃ and 21.7℃ respectively, 1.1℃ less than the above group. Therefore, we can deduce that the white bandage is not prevent radiation fully, and the deviation may be more obvious than we thought before.

3.3. Result-3 (Wind Prevention Analysis)

The deviation caused by the wind is tested in the experiment and the data is shown in figure 5. In the morning, all the lines in the diagram are almost in the same level which is because the radiation impact is more obvious than wind so that the influence of the wind could be ignored. However, when the solar radiation abated gradually, the influence of wind on the experiment increased, especially at night. For instance, line-11, line-12 and line-13 are 3℃ degree lower than the others due to the lack of air ventilation. Because the indoor air temperature is about 20℃ which is higher than the windowsill surface, we can infer that the white bandage group without transparent adhesive tape (line-4, line-5 and line-6) has been heated through the way of cross-ventilation.

3.4. Result-4 (thermal isolation analysis)

Figure 6 shows the experiment data which is collected through the iButton that fixed by the white cotton bandage which have the excellent performance in heat preservation and insulation. These bandages are in different mount of layers (2, 4, and 6 respectively) and the differences among the three test are not evident. We assumed this phenomenon is due to the small span of the layers' quantity, which will be verified in later experiments.

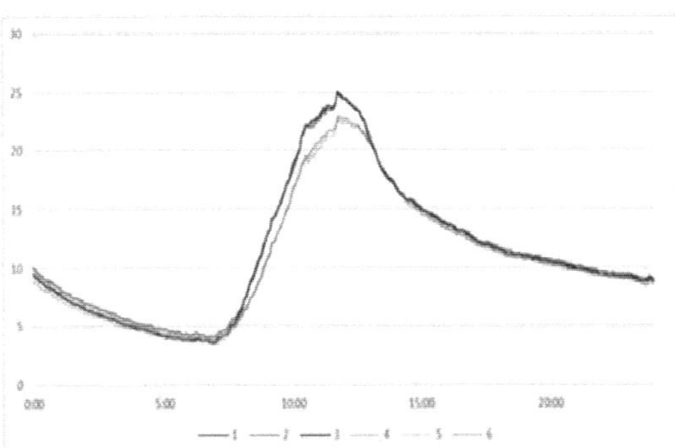

Figure 3. bandage color analysis.

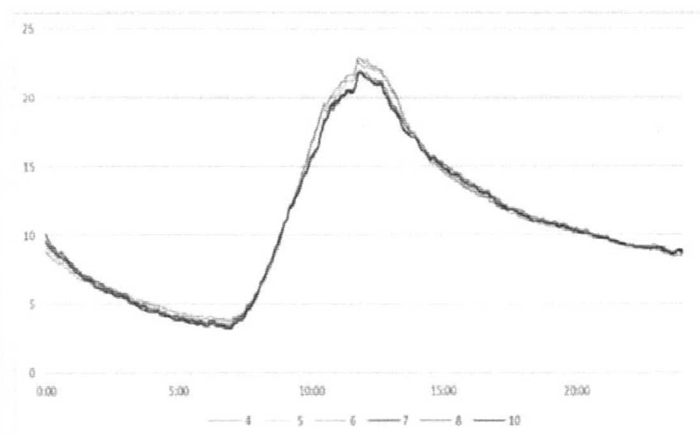

Figure 4. radiant prevention analysis.

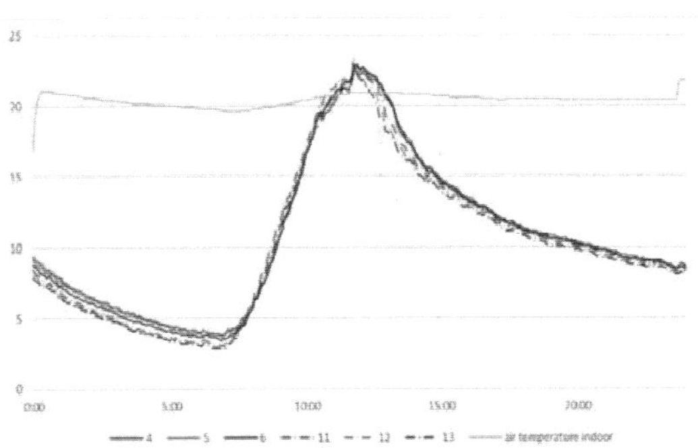

Figure 5. wind prevention analysis.

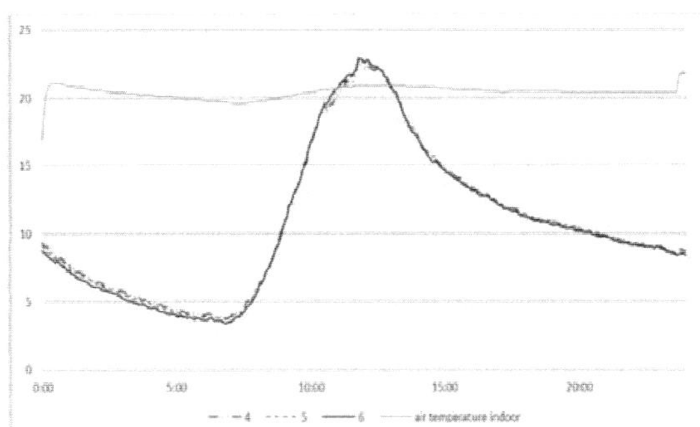

Figure 6. thermal isolation analysis.

4. Conclusion

This study used the comparative tests to analysis the devices' accuracy that could be affected especially when these thermometers have been fixed by the different kinds of material. Black materials can make the value significantly higher than the real temperature. However, when we add aluminum foil paper which could eliminate the radiation on the surface, the temperature value declined. In addition, windproof material can prevent the wind effect although the influence of the wind is slighter than solar radiation. Last but not the least, the influence of thermal insulation (bondage layers) on iButton's temperature measuring was not obvious in this study.

References

[1] Roznik E A and Alford R A 2012 *J. Therm. Biol.* Does waterproofing Thermochron iButton data loggers influence temperature readings? **37(4)** 260-4.

[2] van Marken Lichtenbelt W D, Daanen H A, Wouters L, Fronczek R, Raymann R J, Severens N M and Van Someren E J 2006 *Physiol. Behav.* Evaluation of wireless determination of skin temperature using iButtons, **88(4-5)** 489-97.

[3] Zhang H, Arens E, Huizenga C and Han T Y 2010 *Build. Environ.* Thermal sensation and comfort models for non-uniform and transient environments: Part1: Local sensation of individual body parts, **45(2)** 380-8.

[4] Cosma A C and Simha R 2018 *Build. Environ.* Thermal comfort modeling in transient conditions using real-time local body temperature extraction with a thermographic camera, **143** 36-47.

[5] Califano R, Naddeo A and Vink P 2017 *Appl. Ergon.* the effect of human-mattress interface's temperature on perceived thermal comfort, **58** 334-41.

[6] Smith A D H, Crabtree D R, Bilzon J L J and Walsh N P 2010 *Physiol. Meas.* The validity of wireless iButtons and thermistors for human skin temperature measurement, **31** 95-114.

[7] ISO-standard 9886 Ergonomics – Evaluation of Thermal Strain by Physio- Logical Measurements International Standards Organization, Geneva, 2004.

[8] Schellen L, Loomans M G L C, de Wit M H, Olesen B W and van Marken W D 2012 *Physiol. Behav.* Lichtenbelt The influence of local effects on thermal sensation under non-uniform environmental conditions—gender differences in thermophysiology, thermal comfort and productivity during convective and radiant cooling, **107** 252-61.

[9] Song C, Liu Y, Zhou X, Wang X, Li J and Liu J 2016 *Build. Environ.* Temperature field of bed climate and thermal comfort assessment based on local thermal sensations, **95** 381-90.

[10] Rubinstein E H and Sessler D I 1990 *Anesthesiology* Skin-surface temperature gradients correlate with fingertip blood flow in humans, **73(3)** 541-5.

[11] Angilletta M and Krochmal A R 2003 *Herpetol. Rev.* The Thermochron: a truly miniature and inexpensive temperature-logger, **34(1)** 31-2.

[12] Dechmann D K N, Kalko E K V and Kerth G 2004 *Evol. Ecol. Res.* Ecology of an exceptional roost: energetic benefits could explain why the bat Lophostoma silvicolum roosts in active termite nests, **6** 1037-50.

[13] Isaac J, De Gabriel J L and Goodman B 2008 *Anim. Conserv.* Microclimate of daytime den sites in a tropical possum: implications for the conservation of tropical arboreal marsupials, **11(4)** 281-7.

Acknowledgements

The research content of the thesis comes from the Chinese postdoctoral special funded project "Research on the dynamic prediction and evaluation of Harbin human settlement environment integrating CA and WRF models" (2017T100244).

FEM strategies for Large Scale Thermo-Mechanical Simulations with Material Non-linearity

Francesco De Bona[1,*], Denis Benasciutti[2], Luciano Moro[1] and Jelena Srnec Novak[1]

[1]DPIA, University of Udine, Italy

[2]DE, University of Ferrara, Italy

debona@uniud.it, bnsdns@unife.it, luciano.moro@uniud.it, jelena.srnec@uniud.it

Abstract. Numerical techniques based on the Finite Element Method (FEM) are mature to provide computational tools that permit multi-physical problems to be dealt with. A relevant case concerns thermo-mechanical simulations of industrial components and processes. Nevertheless, a thermo-mechanical FEM model generally requires a transient non-linear analysis where the variation of material properties with temperature, as well as plasticity and creep, have to be considered. It follows that large scale models are often obtained and unfeasible computational time is thus required. The aim of this work is to put in evidence the possible scale reduction that can be achieved introducing model simplifications based on a practical engineering approach.

1. Introduction

The Finite Element Method is a powerful tool to investigate the thermo-mechanical behavior of industrial components and processes. The main computational aspects have been examined in depth [1] and several commercial codes are available (i.e. see [2]). Nevertheless, the scientific literature still lacks of practical guidelines to deal with large-scale problems. In fact, such simulations, as a matter of principle, require a transient analysis of a three-dimensional model, where several non-linearities have to be taken into account (variation of the material properties with temperature, plasticity and creep) [3]. It follows that, quite often, unfeasible computational time is required. The aim of this work is to present practical modelling strategies suitable to reduce significantly the scale of the problem.

A few representative examples are here discussed. It is shown that in certain circumstances a transient analysis can be replaced with a steady state simulation. Nevertheless, if a transient analysis cannot be avoided, a huge reduction of the problem dimensionality can sometimes be achieved by referring to harmonic models or, when cyclic plasticity has to be considered, by applying a useful acceleration technique based on an appropriate correction of the stabilization speed in a combined non-linear kinematic and isotropic model.

2. Transient or Steady State Analysis

A thermo-mechanical simulation generally consists of a transient thermal analysis followed by a mechanical analysis. Considering the non-linearities due to the variation of material properties with temperature (conductivity, modulus of elasticity and yield strength), the simulation time can be quite long, especially when the virtual prototyping approach is implemented in the design phase and several simulations, according to different configurations, are required. It is thus a useful practice to check preliminarily if a steady state thermal analysis is able to give acceptable results. As an example, the case of fire door is here presented. Such component must comply with product certification requirements, which usually need a fire test, where one side of the component is heated by means of a prescribed temperature-time relationship. Throughout the duration of the test, the component must fulfill specific thermal and mechanical requirements. In particular, the mean temperature on the cold side (not exposed to the fire) should not exceed a prescribed value and the maximum gap between panels and frame should not allow flame and/or smoke to propagate. Recently it has been observed [4, 5], that an uncoupled thermo-mechanical FEM analysis is able to assess accurately the distortion induced by the thermal loads.

A fire door is constituted by two thin steel plates with an insulating material (e.g. rock wool) in between, which is able to act as a thermal barrier that prevents heat flow.

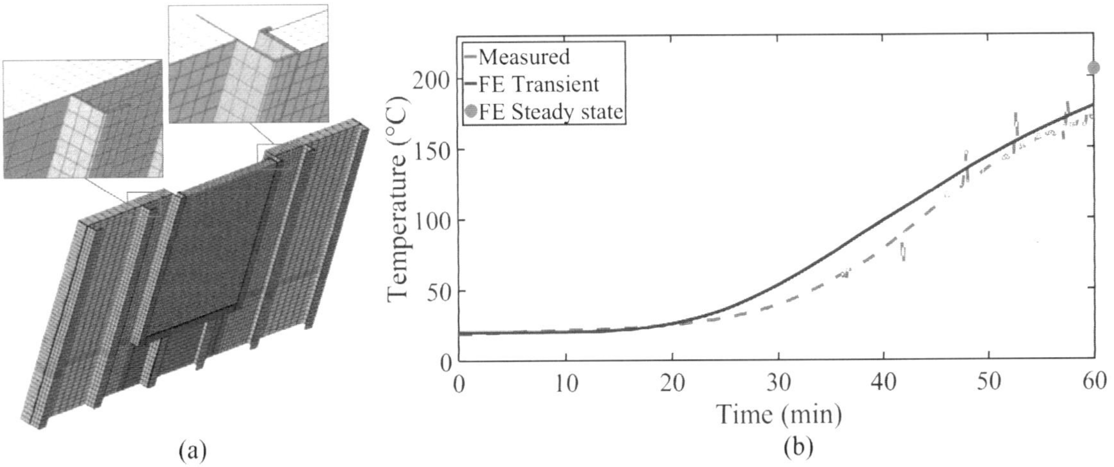

Figure 1. FEM simulation: fire door FEM model (a) comparison between the steady state and the transient cases with measurements (b).

Figure 1a presents the FEM model (halved geometry, due to the symmetry with respect to a meridian horizontal plane) with a significant number of element. In figure 1b, the temperature on the cold side of the door is shown according to a transient and a steady state analysis. The comparison with measurements shows that the approximate solution gives quite satisfactory results. This is due to the fact that, in this case, the temperature variation in the oven is quite slow and, therefore, a negligible error is introduced by imposing directly the oven temperature to the hot side of the door.

3. Symmetry, Axi-Symmetry and Semi-Analytical FEM

In FEM modelling a well-known and widely used approach to reduce the dimension of the problem is that of exploiting any symmetry, if present. Referring to the thermo-mechanical analysis, in [6], the double symmetry along the longitudinal axis allows the behavior of a squared mold for continuous casting to be modeled by considering only one fourth of the geometry. In the case of an anode for electric arc furnace [7], the evaluation of the stress and strain induced by a thermal cycling requires also a phase change transition to be modelled; in fact, very high temperatures can be reached and it may happen that some areas undergo a partial melting. It follows that the model complexity (and consequently the computational time) strongly benefits from adopting a plane axi-symmetric model.

A less obvious and more interesting case of scale reduction is that of a work roll in hot rolling plants. If the study is not addressed to the process, but to the component, and if a durability analysis has to be carried out, a transient thermal analysis (material conductivity varies with temperature), followed by a non-linear mechanical analysis (Von Mises plasticity with kinematic hardening and material properties varying with temperature) has to be performed. Even if a plane model is adopted [8], unexpectedly, a huge computational time is required, in particular to perform the mechanical analysis. By observing that the work roll has an axi-symmetric geometry and is non-axisymmetric loaded (thermal flux coming from the billet and water cooling acting non-axi-symmetrically), a semi-analytical method [9] based on 1D harmonic finite elements can be adopted. Although such an approach is generally used to address three-dimensional problems, in this case it is used to make one-dimensional a two-dimensional case, as shown in figure 2. Adopting a plane model, a transient time of 43 seconds was simulated in 10 days. On the contrary, the harmonic one-dimensional model permits the mechanical behavior over a larger time interval to be easily simulated with reasonable computational times (3-4 hours).

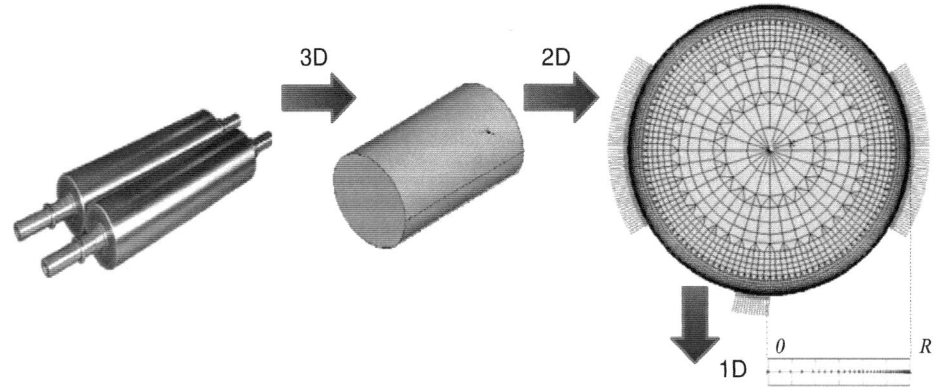

Figure 2. Working roll simplified model: from 3D to 1D.

4. Acceleration of the Simulation by Modifying Material Parameters

Often thermo-mechanical analysis is performed in the framework of a durability analysis [10], in particular when cyclic thermal loads are sufficiently high to produce plasticization in some areas of the mechanical component under investigation. In this case, it has been observed [11, 12] that the cyclic plasticity material model has to be defined accurately, generally making use of combined non-linear kinematic and non-linear isotropic models (monotonic hardening was simulated with a Chaboche model, using three pairs of C and parameters, while cyclic softening phenomena was described with Voce approach). Variation of material parameters with temperature was also taken into account. Typical case is that of a mold for continuous casting of steel, which provides an initial solidification of the steel product. Molds are generally constituted by a pipe made of copper alloy and surrounded by a steel jacket. The cooling water circulates in the cavity between these two parts. Due to the presence of molten steel, the mold undergoes a high thermal flux and consequently high thermal gradients and stresses exceeding yielding are produced (see figure 3a). Moreover, during service, the component is subjected to fluctuating thermal loads because of planned switch-offs for maintenance. Such loads can lead thermal cracks to nucleate, thus reducing the component service life [13, 14]. A thermal distortion can be also observed [15, 16].

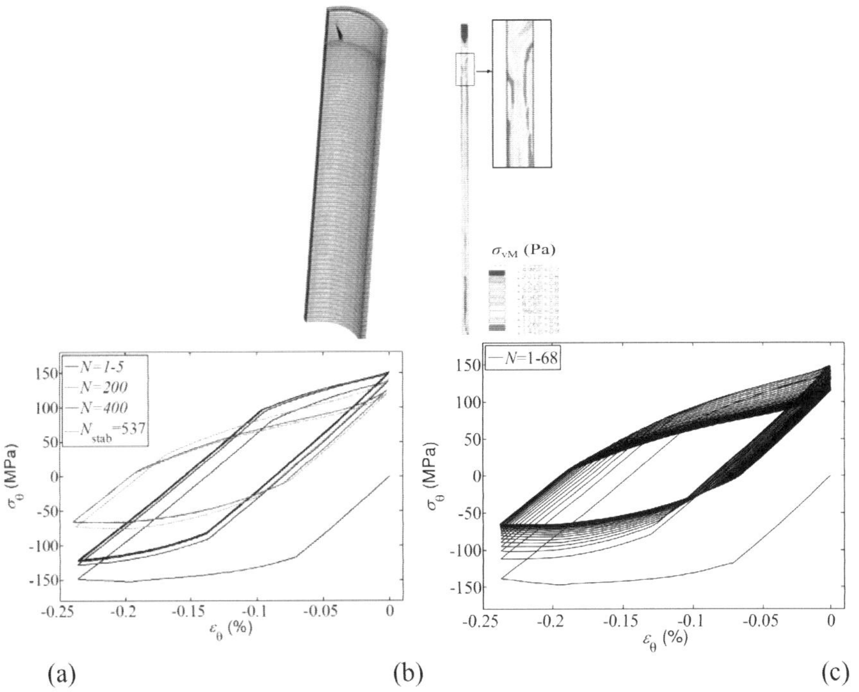

Figure 3. FEM model and von Mises stress distribution (a), evolution of the hoop stress-strain curves: combined (b) and accelerated (c) models.

To perform a durability analysis, it is necessary to estimate accurately the stress-strain evolution caused by the fluctuating thermal flux up to stabilization. In certain circumstances (material characteristics, amount of plastic strain range induced by thermal loads, etc.), the number of cycles to reach stabilization could be very high, and therefore the simulation could be unfeasible. An acceleration technique has been proposed in [17, 18] to speed-up the analysis, this approach being based on a fictitious increment of the parameter governing the speed of stabilization in the isotropic model. Figure 3b displays the stress-strain evolution that occurs in the most stressed part of the mold. It can be observed that stabilization is reached after 537 cycles, which means that a huge number of heating and cooling cycles have to be simulated. Figure 3c shows the same case, where the isotropic model was corrected by increasing the parameter controlling the speed of stabilization, in order to reach stabilized conditions only after 68 cycles. This approach permits a strong reduction of the computational time without affecting significantly the equivalent strain range in stabilized condition and therefore permitting a correct durability analysis to be subsequently performed [19].

5. Discussion and Conclusions

In this work several engineering applications requiring a thermo-mechanical FEM simulation were discussed. Although such models generally imply a very high computational time, possible strategies to shorten the analysis, without affecting significantly the results, were presented. Tab. 1 summarizes the results; in the presented cases it can be observed that a drastic reduction of the computational time, up to 80%, is always achieved. In particular, substituting a transient with a steady state analysis permits a 95% reduction of the computational time. When a durability analysis has to be performed and the component is subjected to thermal cycles which induce plasticity and, therefore, a thermal fatigue analysis has to be performed, acceleration techniques based on combined non-linear kinematic and isotropic models seem providing quite promising results (88%-time reduction). Even higher scale reduction (98%) can be achieved by using harmonic elements.

Table 1. Comparison between the presented cases.

Case	Type of model-analysis	Computational time	Time reduction (%)
Fire test	Transient Steady state	20 min 1 min	95
Roll mill	Plane model 1D Harmonic model	10 days 4 hours	98
Mold	Combined model Accelerated model	108 min 14 min	88

References

[1] Bergheau J M 2014 Thermomechanical Industrial Processes: Modeling and Numerical Simulation, ISTE Ltd, London, UK, John Wiley & Sons, Inc, Hobochen (NY) Eds.

[2] Regener B, Krempaszky C, Werner E and Stockinger M 2011 Thermo-mechanical FE2 simulation scheme for Abaqus, PAMM **11(1)** 547-8.

[3] Nishikawa H, Serizawa H and Murakawa H 2007 *Sci. Technol. Weld. Joi.* Actual application of FEM to analysis of large scale mechanical problems in welding, **12(2)** 147-52.

[4] Boscariol P, De Bona F, Gasparetto A and Moro L 2015 *J. Fire Sci.* Thermo-mechanical analysis of a fire door for naval applications, **33(2)** 142-56.

[5] Moro L, Boscariol P, De Bona F, Gasparetto A and Novak J S 2017 *Fire Technol.* Innovative design of fire doors: computational modeling and experimental validation, **53(5)** 1833-46.

[6] Novak J S, Stanojevic A, Benasciutti D, Bona F D and Huter P 2015 *Procedia Eng.* Thermo-mechanical finite element simulation and fatigue life assessment of a copper mould for continuous casting of steel, **133** 688-97.

[7] Moro L, Benasciutti D and Bona F D 2019 *Ironmak. Steelmak.* Simplified numerical approach for the thermo-mechanical analysis of steelmaking components under cyclic loading: an anode for electric arc furnace, **46(1)** 56-65.

[8] Benasciutti D, Bona F D and Munteanu M G 2016 *Strain Anal. Eng. Des.* A harmonic one-dimensional element for non-linear thermo-mechanical analysis of axisymmetric structures under asymmetric loads: the case of hot strip rolling, **51(7)** 518-31.

[9] Munteanu M G, Bona F D and Bressan F 2018 *Mech. Based Des. Struc.* Shaft design: a semi-analytical finite element approach, **46(2)** 184-95.

[10] Manson S S and Halford G R 2006 *ASM Int.* Fatigue and Durability of Structural Materials, #06987G.

[11] Benasciutti D, Novak J S, Moro L and Bona F D 2018 *Fatigue Fract. Eng. Mater. Struct.* Experimental characterisation of a CuAg alloy for thermo-mechanical applications. Part 2: Design strain-life curves estimated via statistical analysis, **41(6)** 1378-88.

[12] Novak J S, Benasciutti D, Bona F B, Stanojevic A, Luca A D and Raffaglio Y 2016 *IOP Conf. Series: Mater. Sci. Eng.* Estimation of material parameters in nonlinear hardening plasticity models and strain life curves for CuAg alloy, **119(1)** 012020.

[13] Park J K, Thomas B G, Samarasekera I V and Yoon U S 2002 *Metall. Mater. Trans. B* Thermal and mechanical behavior of copper molds during thin-slab casting (II): mold crack formation, **33(3)** 437-49.

[14] Novak J S, Lanzutti A, Benasciutti D, Bona F D, Moro L and Luca A D 2018 *Eng. Fail. Anal.* On the damage mechanisms in a continuous casting mold: After-service material characterization and finite element simulation, **94** 480-92.

[15] Moro L, Novak J S, Benasciutti D and Bona F D 2018 *Key Eng. Mater.* Copper Mold for Continuous Casting of Steel: Modelling Strategies to Assess Thermal Distortion and Durability, **754** 287-90.

[16] Moro L, Novak J S, Benasciutti D and Bona F D 2019 *Ironmak. Steelmak.* Thermal distortion in copper moulds for continuous casting of steel: numerical study on creep and plasticity effect, **46(1)** 97-103.
[17] Novak J S, Moro L, Benasciutti D and Bona F D 2018 *Proc. Struct. Integr.* Accelerated cyclic plasticity models for FEM analysis of steelmaking components under thermal loads, **8** 174-83.
[18] Novak J S, Bona F D, Benasciutti D and Moro L 2018 *MATEC Web Conf.* Acceleration techniques for the numerical simulation of the cyclic plasticity behaviour of mechanical components under thermal loads, **165** 19010.
[19] Benasciutti D, Novak J S, Moro L, Bona F D and Stanojević A 2018 *Fatigue Fract. Eng. Mater. Struct.* Experimental characterisation of a CuAg alloy for thermo-mechanical applications. Part 1: Identifying parameters of non-linear plasticity models, **41(6)** 1364-77.

The Simulation Model of the Resistance Factors Affecting the Adoption of Healthcare Robot Technology in Tertiary Care Center of Thailand

Paniti Vichitkraivin[1] and Thanakorn Naenna[1]

[1]Department of Industrial Engineering, Faculty of Engineering, Mahidol University, Thailand

Paniti.v.z.2535@gmail.com, thanakorn.nae@mahidol.ac.th

Abstract. Healthcare robot technology has become the important tool and it has extremely important role in the healthcare field. In spite of it great potential, gaps exist in our grasp of how users will assess change related to the healthcare robot. Previously research shows that when the new technology is used, the user may make decision to accept or resist technology. According to barrier theories this study develops a modify resistance model to describe the factors which resist the medical staffs' intention to use healthcare robot. The participants were selected from medical staffs (physician, nurse, and staffs) in Tertiary care center located in Thailand. We collected data using questionnaire to test a research model. The research model is analyzed by confirmatory factor analysis (CFA). The empirical results confirm that technological barrier, time barrier and change barrier have negative effect on medicals' behavioral intention to use healthcare robots. This study is discussed several practical and theoretical implications.

1. Introduction

Healthcare technology is developed and researched widely in healthcare service at present [1] such healthcare information system. When healthcare technology is developed alongside the need of patient are increasing, the healthcare robot has used in healthcare service. Healthcare robots initiative have been promise broadly [2]. They have become the important tool in healthcare service and they have had important role in healthcare since they can improve the healthcare service quality [3, 4]. Moreover, the healthcare robot market is growing fast and the highest value per unit sold [5].

Although healthcare robot can reduce healthcare cost and error from medical staffs, however the success of the healthcare robot initiative is dependent not only on robot designers, but also on medical staffs' willingness to adopt these healthcare robots because medical staffs are the user of healthcare robot to provide services to patients. Corporate executives or decision makers and robot designers, so, need an understanding of the adoption and barrier factors which have affected facilitate the adoption and using of healthcare robot by medical staffs [6].

The healthcare robot literature has focused on the technical using and design of healthcare robot [7-9], the adoption study area is the new topic in healthcare service. Previous researches on healthcare robot adoption have applied the technology acceptance model for suitable healthcare robot context, they have recommended the critical factor of healthcare robot adoption such as Alaiad A and Zhou L research [3]. However, user resistance is unavoidable for management and it may result in lower performance than expected. As a result, organizations often get less of performance than they expect when they invest in new technology [10]. So, there is a need to check the critical resistance factors that affect technology adoption.

The previous researches on healthcare robot usage have been limited and most of them don't care problem about the resistance factors of users [11]. So, this research is the determinants of resistance factors which effect healthcare robot adoption of medical staffs. This study was motivated by observation that extant theories of technology usage, such as Unified Theory of Acceptance and Use Technology (UTAUT) have focused almost exclusively on users' positive (enabling) perceptions in relation to technology usage while ignoring negative (inhibiting) perceptions that may prevent technology usage, according to Cenfetelli's study [12]. Although Cenfetelli did not identify specific inhibitors of technology usage. Based on our literature review, resistance factors model of this research is created by using of resistance theory is main and adaptation with obstacles factors for suitable healthcare robot context.

2. Background and Theoretical

2.1. Healthcare Technology

Now healthcare institutes attempt to improve their services and increase the quality as much as possible. The quality improving in the hospitals can't be accomplished without healthcare technologies using and the other facilities since healthcare technologies have the potential to reduce diagnosis error, reduce process time and increase safety and quality. So, technology is playing an important role in the healthcare industry. Before, healthcare institutes began installing information and communication technology system in various ways; from the very simple format as using basic computer software putting the data in till the complex format as using healthcare robot [6] which can reduce 100% error [13]. The usage express IT in the healthcare section is very important to create competitive advantage and increase business value [6].

2.2. Robot

Now, robots have been widely used such as automotive industry (this robot is used in automotive production), researching industry (this robot cans co-work on scientific research and surveying with human) and healthcare industry (this robot can use mechanical-armed robot for taking part in the operation because it is able to do accuracy required work better than human can do) since each company wants to reduce labor costs and human error. Though the definition of robot is currently no clear for adoption by the general public and there is no unanimous opinion that what qualified to be a robot [14]. Each country has the difference robot standard for consideration. So, to understanding that match about the Thailand standard of robot, the robot standard in this research is identified by multiple theories. First, robot is the machine that is able to work, some of them can work by itself meanwhile the others need instructor, according to NASA [3, 6]; second, robot should be able to do one or more of the following: 2.1. move around 2.2. operate a mechanical limb 2.3. sense and manipulate their environment 2.4. exhibit intelligent behavior, following from Thailand Center of Excellence for Life Sciences (TCELS), [14]; and third, robot is the mechanical tools and mechanisms inside can be many things together with a human or a human agent and can rearrange work process, according to Saranukromthai book 36 [15].

2.3. Healthcare Robot

Robot is the machine which can be used in the work. Some of the robots can work by themselves but the other robots need controller, the command must be entered or programmed to make it work, according to NASA [3, 6]. When these tasks mean to healthcare, these robots are called healthcare robot. Healthcare robots have an important role in improving the healthcare system and assistance the medical staff in every stage of the healthcare process [4]. They include drug manufacturing, nursing, drug medication management, dispensing drugs to patients, monitoring patient vital signs and carrying out surgeries and telemedicine [16]. Healthcare robots have been recently used by Thailand hospital such as Siriraj hospital, Bangkok and Ramathibodi hospital, Bangkok. According to the literature, Thailand healthcare robot can be dividing into three major categories include: treatment robot, support robot and medical service robot [4]. This classification of robot has difference applications and tasks characteristics depend on the role of robots and the appropriateness of the responsible service.

The Healthcare robot in Thailand hospital include surgery robot, telemedicine robot, nurse healthcare robot, pharmacy robots, physical robot, CT scan, MRI and specific healthcare robot such as robot helps develop children. These healthcare robots are considered in this research and they have characteristics that match with robot standard.

2.4. Technology Acceptance Barrier for Healthcare Robot

The resistance or barrier factor is the reason that humans don't show any behavior. The diffusion research theory barrier has related to explaining the different between innovators and early adopters and other ideal types of adoption behavior, such as the early and late majority and laggards [17]. These barriers are closely associated to all type problem issues. Each barrier factors have very significant for the problem of equipment used in every work which make technology acceptance barrier is specified into various methods and theories is suitable for work. Now, there are many barrier research theorists such as: Pagani (2004) [18] and Kleijnen et al. (2004) [19].

From literature review, this research is applying the Pagani barrier model because it accentuates to the main in characteristics of the technology and most of its factors is suitable of healthcare robot context. These barrier factors include: technological barrier ((Lack of) Ease of use Barrier factor and Functional Usability Barrier factor), security barrier, Time barrier (Long time training and setting barrier) and change barrier.

2.5. Technology Acceptance and Resistance

When the new technology is used, the user may make decision to accept or resist technology. It depends on assessing changes related to the new system [20]. Although healthcare technology has great potential to improve the healthcare quality but this benefit may not receive the attention. Due to many healthcare technology efforts are experiencing problems or failures. Several failures, and these issues can be checked from the resistance of the user [21]. Bhattacherjee A and Hikmet N said the resistance isn't equivalent to not use because not using may mean that the users who use that technology don't know about new technology or continue to assess the technology before adoption. But resistance means that technology was considered and was rejected by users [22]. However, the ability to accept and resistance the technology must be examined in conjunction with the theoretical model of the common due to the resistance of the users is an important barrier to use new technology [12]. So, this research determined the resistance of the user. It is the conflict of the user to the changes associated with the use of new technologies.

Hsieh and Pi-Jung research showed barrier factors of acceptance healthcare technology have negative effected with behavioral intention to use [11, 23].

This research aims to answer the following the two mains research question: 1) What are the barrier factors influencing the medical staffs' behavioral intention to use healthcare robots? 2) What is the barrier factors which have strong effect of healthcare robot adoption?

3. Research Model and Hypotheses

In this study, the research model is presented in figure 1. The model base on Pagani barrier model and result of Hsieh and Pi-Jung research to examine the barrier factors effect of medical staffs' behavior toward adopting a healthcare robot. The barrier factors in this research model include technological barrier ((lack of) ease of use barrier factor and functional usability barrier factor), security barrier, time barrier (long time training and setting barrier), change barrier.

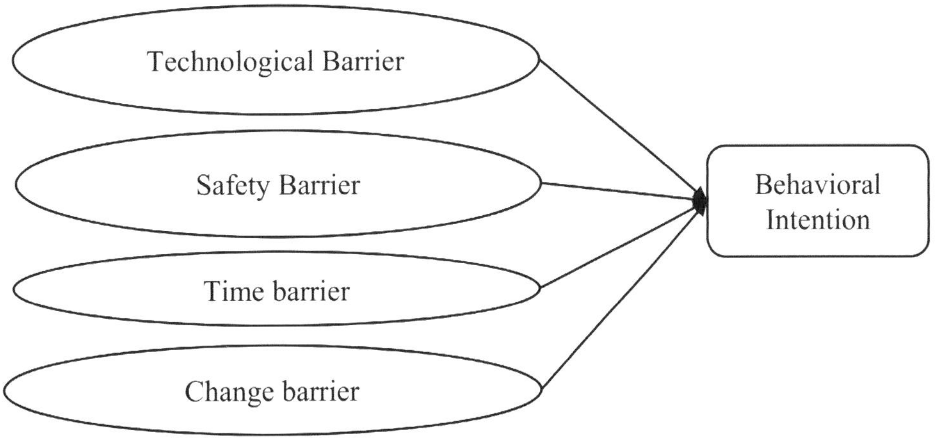

Figure 1. The barrier model in this research.

3.1. Technological Barrier ((Lack of) Ease of Use and Functional Usability Barrier Factor)
(Lack of) ease to use barriers means the each person's feeling that the technology was difficult for using when he must actual use. It was not flexible and study hard. There is the research show lack of ease of use causes medical staff refusing of using healthcare robot [24, 25]. In this research means the medical staffs found healthcare robot so complicated, difficult to use when it needs actually work and learning is difficult so they didn't want to use.

Functional usability barrier means the feelings of users that technology can't help their job effectively enough by itself or technology isn't response their work. The result of healthcare robot research show the lack of functions using is one of the barrier which causes the medical staffs don't accept using the healthcare robot [24, 25]. For this research means the medical staffs' feeling that healthcare robot can't response their job.

H1. Technological Barrier has a negative effect on users' behavioral intention.

3.2. Safety Barrier
Safety barrier means the users' feeling that using technology won't safe for users and patients. Safety barrier has influential for healthcare robot acceptance by the relevant concept, they have a wide range of ideas such as: BenMessaoud C. and group research (2011), Chen et al. (2007) and Stahl & Coeckelbergh, (2016) [24, 26, 27]. This research context means the medical staffs' feeling to use healthcare robot that isn't safe for them and patient when there is actual using.

H2. Safety Barrier has a negative effect on users' behavioral intention.

3.3. Time Barrier (Long Time Training and Setting Barrier)
Long time training barrier means the users' feeling that the time in training of using technology is spent too much, they feel waste of time in their work and bored with. Previous research results show that the long time too much training is the barrier that make the medical staffs against the adoption of healthcare technology. The researches support that content are follows Alaboudi et al., 2016, Treister, 1998 and BenMessaoud et al., 2011 [24, 25]. For this research context means the medical staffs' feeling, the training for using healthcare robot spends too much time and makes them wasting time and boredom.

Long time setting barrier means the users' feeling that they spend too much time for setting technology. BenMessaoud C. and group research said the spending too much time for setting surgery robot is the barrier factor of healthcare robot the because medical staffs will afraid that they can't help patient in time [24]. In this research context means the medical staffs feel that healthcare robot must spend too much time for setting so they can't help patient in time.

H3. Time Barrier has a negative effect on users' behavioral intention.

3.4. Change Barrier

Change barrier means the users don't want to change the traditional healthcare in order to use new healthcare technology because of personal reasons. Previous researches show the change barrier is the result from several human causes such as the lack of clear benefits, unknown technology phobia and not necessary to change. Many researchers said that the change is barrier of technology acceptance such as Panteleimonitis S and group (2017) [28], Marescaux J (2005) [29]. In this research means the medical staffs don't want to change their traditional healthcare to using healthcare robot in actual working because personal reason.

H4. Change Barrier has a negative effect on users' behavioral intention.

4. Methodology

This study focuses on medical staffs who use healthcare robot include physicians, nurses, and staffs. The participants were selected form Tirtiary Care Center local in Thailand. To help participants understand about healthcare robot, we explain about robot standard, functions, and characteristics of healthcare robots in the introduction of the survey. We collected 225 responses, 189 returns and among which 171 were valid and used in this study and 18 were incomplete. Among the participants, 44.44% were male and 55.56% were female.

After creation the research model based on the sources mentioned earlier, we survey to collect the data about barrier factors effecting of medical staffs' perceptions of healthcare robots for testing the research model. The model is tested by using Lisrel program. The survey was conducted by using questionnaire. All the questions were adapted to the suitable of healthcare robot context, and measured on 5-point Likert scales.

5. Result

A reliability analysis was proceeded by using Cronbach's Alpha with 1-5 Likert scales. To test for validity, Item-Objective Congruence (IOC) testing technics has been done. All of scales of Cronbach's Alpha have nearly or exceed 0.70 for each factors group and for overall questionnaires. All questions get the IOC score greater than 0.50 which are declared to be "passed", according to the IOC concepts.

In this research, the confirmatory factor analysis (CFA) was taken to test the hypotheses and check the loading of barrier factors. First measurement model is analyzed; all data observed of latent variables are checked and specified the loading of each measurement items. Theoretically, the barrier factor of healthcare robot adoption by medical staffs in tertiary care center of Thailand should have 4 dimensions include technological barrier, safety barrier, time barrier and change barrier. But from the analysis of question, it was found that barrier factors consisting of technological barrier, time barrier and change barrier in latent variables have been accept, they have t-value more 1.96 (chi-square/df<2, GFI AGFI TLI NFI >0.95 and RMSEA<0.05).

The CFA was taken again to find the hypotheses and barrier factors loading for each barrier factor that have been accepted, the model for healthcare robot with barrier factor loading is shown in figure 2. Acceptable barrier factor items have been found to be negatively associated with behavioral intention to use healthcare robot, result of the factor analysis reported in table 1.

Table 1. Result of the 2nd order confirmatory factor analysis.

Factor	Factor loading	T-value	Percentage Approximation (%)	Result
Technological Barrier	0.92	3.29	32.5	Accept
Time Barrier	0.93	7.46	32.87	Accept
Change Barrier	0.98	9.19	34.63	Accept

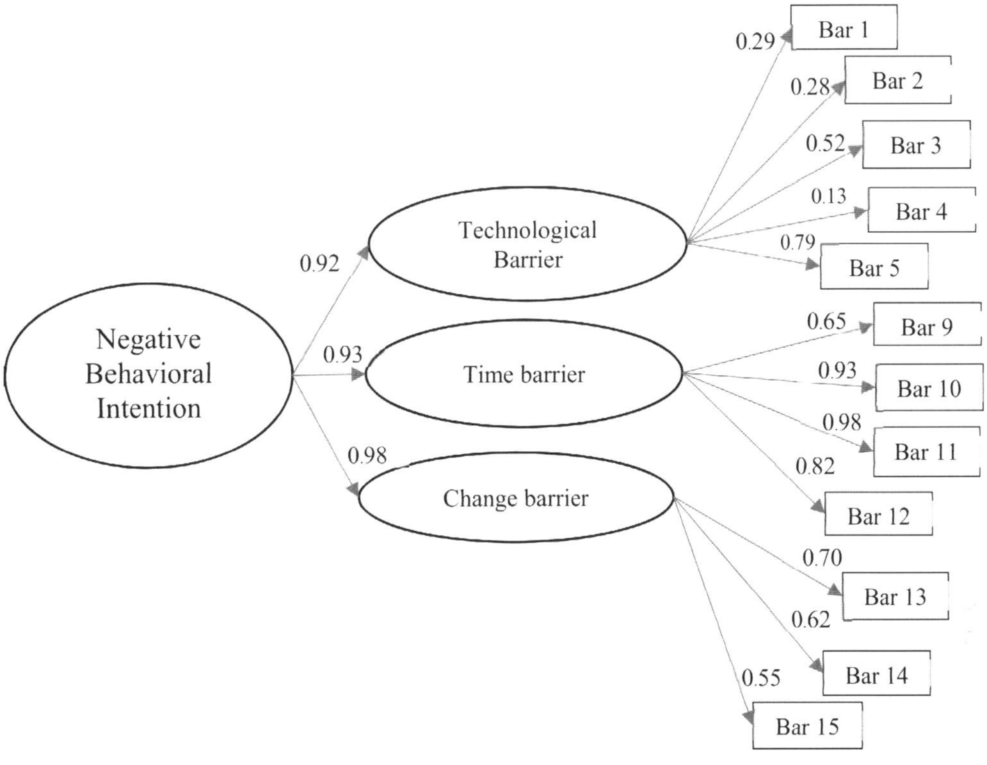

Figure 2. Result of barrier factor loading model.

The change barrier factor has the highest effect to resist of healthcare robot adoption, following with time barrier. The technological barrier has the least influence on healthcare robot adoption.

6. Discussion

In this study empirical study, we analyzed the medical staffs' resistance to healthcare robot adoption in tertiary care center of Thailand. This research model is analyzed by using the confirmatory factor analysis (CFA) for confirmation hypothesis and loading factors. Our study confirmed that medical staffs' resistance to healthcare robot adoption was caused by technological barrier, time barrier and change barrier. The change barrier has highest loading factor caused by the medical staffs don't know the clear benefits enough of robot, not necessary to change and uneasy due to fear. Previous research result shows the solutions include: first, offering strong, committed and positive leadership to lead the implementation process, this team should have implementation leader, engineers, training supervisors, and client support personnel [30] and second, education and training sessions to instruct the users [30]. The time barrier is the second loading factor because the training and preparing robot spend too much time. Resulting in the staff being bored, wasting time and not wanting to use. The solution is Healthcare robot should be designed the function to be similar to the original for reduce training and setting time [24, 31, 32]. The last barrier is technological barrier because some healthcare robot is hard both using and training, unable response their work, no staff teaches how to use and the result are not clear. Previous research suggested the solutions as follow: first, healthcare robot should design to be easy to use and to learn [24]; second, they need to be designed usage condition to be similar to the original [31, 32]; third, their function must response the work and better the original [24] and testing software performance and quality, prior to the using [33]. The safety barrier has influenced only some healthcare robot such as surgery robot [24, 29].

In practical, we consult with actual users in Thailand about the solution of barrier factors in healthcare robot adoption context. They suggest the solutions of these barrier as follows: first, healthcare robot must easy to use; second, the hospital should have training for using healthcare robot by expert in suitable time; third, healthcare robot must response their job with the precise result; fourth, the hospital need to show the robot advantage and information for reducing staff anxiety and the last, the hospital should have public relations and support for the use of a full range of medical service robots. In pragmatic, the medical staffs said that the safety barrier has affected on executives and decision makers recruiting tools.

References

[1] Nisakorn P and Thanakorn N 2013 *EXCLI* Factors affecting the adoption of healthcare information technology, **12** 413-36.
[2] Natarajan N, Aparna S and Kumar J S J 2012 *Int. J. Comp. Appl.* Robot Aided Remote Medical Assistance System using LabVIEW, **38(2)** 6-10.
[3] Alaiad A and Zhou L 2014 *Int. J. Med. Inform.* The determinants of home healthcare robots' adoption: An empirical investigation, **83(11)** 825-40.
[4] Wongpiromsarn T, Damrongchai N and Vatananan-Thcscnvitz R 2016 *2016 Portland Int. Conf. Manag. Eng. Technol.* Technology development roadmap for medical robotics in Thailand. in Management of Engineering and Technology (PICMET), 3240-8.
[5] Transparency Market Research, Medical Robotic Systems Market Global Industry Analysis, Size, Share, Growth, Trends and Forecast, 2012-2018. 2013. https://www.transparencymarketresearch.com/
[6] Alaiad A and Zhou L 2013 Patients' behavioral intention toward using healthcare robots, in Proceedings of the Nineteenth Americas Conference on Information Systems, Chicago, Illinois.
[7] King C H, et al. 2012 *Disabil. Rehabil. Assist. Technol.* Dusty: an assistive mobile manipulator that retrieves dropped objects for people with motor impairments, **7(2)** 168-79.
[8] Nguyen H, et al. 2013 *2013 IEEE Int. Conf. Robot. Auto.* Ros commander (rosco): Behavior creation for home robots.
[9] Chen T L, et al. 2011 *6th ACM/IEEE Int. Conf. Human-Robot Interact. (HRI)* Touched by a robot: An investigation of subjective responses to robot-initiated touch.
[10] Norzaidi M D, et al. 2008 *J. Comput. Inform. Syst.* A study of intranet usage and resistance in Malaysia's port industry, **49(1)** 37-47.
[11] Hsieh P J 2016 *Comput. Human Behav.* An empirical investigation of patients' acceptance and resistance toward the health cloud: The dual factor perspective, **63** 959-69.
[12] Cenfetelli R T 2004 *J. Association Inform. Syst.* Inhibitors and enablers as dual factor concepts in technology usage, **5(11)** 472-92.
[13] Komchadluek Publisher. Siriraj Hospital develop "pharmacy robot" for speed up the medication and reduce the error. 2017. http://www.komchadluek.net/news/edu-health/265172
[14] Thailand Center of Excellence for Life Sciences (TCELS), Medical Robotics Technology Development Roadmap. 2015, Bangkok: Thailand Center of Excellence for Life Sciences. 79.
[15] Djitt L, et al. 2011 Robot, in Saranukromthai book, **36** 179-225.
[16] Siebert J and Rumiński J 2006 *Int. marit. health* Telemedicine–education and practice, **57(1-4)** 235-42.
[17] Bouwman H, et al. 2007 *Telemat. Informat.* Barriers and drivers in the adoption of current and future mobile services in Finland, **24(2)** 145-60.
[18] Pagani M 2004 *J. Interact. Market.* Determinants of adoption of third generation mobile multimedia services, **18(3)** 46-59.
[19] Kleijnen M, De Ruyter K and Wetzels M 2004 *J. Interact. Market.* Consumer adoption of wireless services: discovering the rules, while playing the game, **18(2)** 51-61.
[20] Joshi K 2005 *J. Inform. Technol. Case Appl. Res.* Understanding user resistance and acceptance during the implementation of an order management system: A case study using the equity implementation model, **7(1)** 6-20.
[21] Bartos C E, Butler B S and Crowley R S 2011 *J. biomed. Inform.* Ranked Levels of Influence model: Selecting influence techniques to minimize IT resistance, **44(3)** 497-504.

[22] Bhattacherjee A and Hikmet N 2007 *40th Annual Hawaii Int. Conf.* Syst. Sci. Physicians' resistance toward healthcare information technologies: a dual-factor model.

[23] Hsieh P J 2015 *Int. j. med. Inform.* Healthcare professionals' use of health clouds: Integrating technology acceptance and status quo bias perspectives, **84**(**7**) 512-23.

[24] BenMessaoud C, Kharrazi H and MacDorman K F 2011 *PloS one* Facilitators and barriers to adopting robotic-assisted surgery: contextualizing the unified theory of acceptance and use of technology, **6**(**1**) e16395.

[25] Alaboudi A, et al. 2016 *2016 IEEE/ACS 13th Int. Conf. Comput. Syst. Appl.* Perceptions and attitudes of clinical staff towards telemedicine acceptance in Saudi Arabia, 1-8.

[26] Chen C C, Wu J and Crandall R E 2007 *Int. J. Electron. Healthc.* Obstacles to the adoption of radio frequency identification technology in the emergency rooms of hospitals, **3**(**2**) 193-207.

[27] Stahl B C and Coeckelbergh M 2016 *Robot. Auton. Syst.* Ethics of healthcare robotics: Towards responsible research and innovation, **86** 152-61.

[28] Panteleimonitis S and Parvaiz A 2017 *Surgery-Oxford Int. Edit.* Robotic colorectal surgery, **35**(**3**) 151-6.

[29] Marescaux J and Rubino F 2005 *Euro. Surgery* Robotic surgery: potentials, barriers, and limitations Roboterchirurgie: Möglichkeiten, Barrieren und Hindernisse, **37**(**5**) 279-83.

[30] Boyter T 2006 Planning Your EHR System: Guidelines for Executive Management, MHCA/SATVA Task Force, USA.

[31] Almutairi B 2011 a strategic roadmap for achieving the potential benefits of electronic health record system in the state of Kuwait, UCL (University College London).

[32] Alateyah S A, Crowder R M and Wills G B 2013 *Int. Conf. Inform. Soc.* identified factors affecting the citizen's intention to adopt e-government in Saudi Arabia, **7**(**8**) 244-52.

[33] Hasanain R A and Cooper H 2014 *J. Health Inform. Develop. Countries* Solutions to overcome technical and social barriers to electronic health records implementation in Saudi public and private hospitals, **8**(**1**) 46-63.

The Numerical Analysis of Screw-type Centrifugal Pumps with Different Rotary Impellers

Wang Yue[1, 2, 3], Liu Jin[1, 2, 3], Xu Juan[1, 2, 3], Liu Gang[1, 2, 3], Zhang Yuanming[1, 2, 3] and Hao Zongrui[1, 2, 3]

[1]Institute of Oceanographic Instrumentation, Qilu University of Technology (Shandong Academy of Sciences), Qingdao, 266100, China
[2]Shandong Provincial Key Laboratory of Marine monitoring instrument equipment technology, Qingdao, 266100, China
[3]National Engineering and Technological Research Center of Marine Monitoring Equipment, Qingdao, 266100, China

wangwill@vip163.com, haozr001@sina.com

Abstract. Screw-type centrifugal pumps with different rotary layers have been numerically simulated. Internal flow fields of Single-rotary impeller and double-rotary impeller are simulated by Reynolds Navier-Stokes function and k-ε viscous model. Stream lines, velocities, and pressure variations are calculated. The stream lines of impellers of two types show that the flow directions are significantly different. The flow distribution inside the pump chamber is approximately the same, and the flow rate is quite different. The pressure distribution of the single-rotary impeller is more uneven than the double-rotary impeller. With different rotary layers, the pressure range has a big difference. The mechanism that the internal flow field of single-rotary impeller pump differs from double-rotary impeller pump rotating layers causes such a large difference in the flow field is discussed.

1. Introduction

The invention of the screw-type centrifugal pump can be traced back to the 1960s. It was first used for port operations such as transporting live fish, and it is the most used fish pump now. It has strong anti-winding and anti-blocking ability and non-destructive ability to the transport medium. Nowadays, the screw-type centrifugal pump expands its range of applications, for example, it has been gradually applied in the fields of environmental protection, papermaking, aviation, etc. [1].

As the application of screw-type centrifugal pumps is continually expanding, the computational fluid dynamics (CFD) method is introduced into the research of the pump. A lot of research and discussion have been carried out by combining the characteristics of the internal flow field, pressure and, vorticity obtained by different calculation methods with the external characteristics and performance curves obtained by experimental means [2-7]. These researches mainly focused on a specific structural impeller of screw-type centrifugal pump, however, there are few works studies the effect of different structural impellers on the pump performance, which may be important in the development of different pumps. In this paper, a single-rotary impeller and a double-rotary impeller are introduced. The characteristics of the internal flow field, such as streamlines, velocities, and pressure of the pumps with the two types of impellers are simulated by CFD method. The mechanism that how different structural impellers influence the characteristics is discussed.

2. Computational Methods

2.1. Control Equations

2.1.1. Governing Equation of Fluid Flow. The laws of mass and momentum are applied for compressive laminar and turbulence flow. All control equations are represented by a conservation law. The coordinate system used to describe the problem is the Cartesian coordinate system in the Euler description. When the flowing fluid is defined as a continuum, the governing equations used to solve are as follows.

Continuity equation (1):

$$\frac{\partial \rho}{\partial t}+\frac{\partial}{\partial x_j}(\rho u_j)=0 \tag{1}$$

u is the instantaneous velocity in the j direction and ρ is the density of the fluid.

Momentum equation (2):

$$\frac{\partial}{\partial t}(\rho u_j)+\frac{\partial}{x_j}(\rho u_i u_j)=-\frac{\partial P}{\partial x_i}+\frac{\partial \tau_{ij}}{\partial x_j}+\rho f_i \tag{2}$$

P is the static pressure, τ_{ij} is the viscous stress tensor and f_i is the volume force. τ_{ij} can be expressed as:

$$\tau_{ij}=\left(\frac{\partial u_i}{\partial x_j}+\frac{\partial u_j}{\partial x_i}\right)-\frac{2}{3}\mu\left(\frac{\partial u_m}{\partial x_m}\right)\delta_{ij} \tag{3}$$

μ is the hydrodynamic viscosity, δ is the Kronecker function.

Bring (2) into (1) to get the Navier-Stokes equation:

$$\frac{\partial}{\partial t}(\rho u_j)+\frac{\partial}{\partial x_j}(\rho u_i u_j)=-\frac{\partial P}{\partial x_j}+\frac{\partial}{\partial x_j}\left(\mu\left(\frac{\partial u_i}{\partial x_j}+\frac{\partial u_j}{\partial x_i}\right)-\frac{2}{3}\mu\left(\frac{\partial u_m}{\partial x_m}\right)\delta_{ij}\right)+\rho f_i \tag{4}$$

2.1.2. Standard k-ε Turbulence Model. The most commonly used model in this study is the standard k-ε turbulence model. To estimate the velocity and length dimension in the standard k-ε turbulence model, turbulent flow energy transfer and dissipation equations are introduced, as shown below:

$$\frac{\partial}{\partial t}(\rho k)+\frac{\partial}{\partial x_j}(\rho u_j k)=\rho P-\rho\varepsilon+\frac{\partial}{\partial x_j}\left(\frac{\mu+\mu_t}{\sigma_k}\frac{\partial k}{\partial x_j}\right) \tag{5}$$

$$\frac{\partial}{\partial t}(\rho\varepsilon)+\frac{\partial}{\partial x_j}(\rho u_j\varepsilon)=C_{\varepsilon_1}\frac{\rho P\varepsilon}{k}-C_{\varepsilon_2}\frac{\rho\varepsilon^2}{k}+\frac{\partial}{\partial x_j}\left(\frac{\mu+\mu_t}{\sigma_\varepsilon}\frac{\partial \varepsilon}{\partial x_j}\right) \tag{6}$$

P is defined by equation (7):

$$P = v_t \left(\frac{\partial u_i}{\partial x_j} + \frac{\partial u_j}{\partial x_i} - \frac{2}{3} \frac{\partial u_m}{\partial x_m} \delta_{ij} \right) \frac{\partial u_i}{\partial x_j} - \frac{2}{3} k \frac{\partial u_m}{\partial x_m} \delta_{ij} \tag{7}$$

2.2. Geometric Model Parameters and Meshing

The parameters of the simulated pumps with single-rotary impeller and double-rotary impeller are volume flow $Q = 80$ m^3/h, head $H = 10$ m and rotating speed $n = 1450$ r/min. The main parameters of the pumps with single-rotary impeller are listed as below.

Table 1. Parameters of pumps with single-rotary impeller and double-rotary impeller.

Parameters	Single-rotary impeller	Double-rotary impeller
Inlet diameter D_1	160[mm]	110[mm]
Outlet diameter D_2	200[mm]	200[mm]
Axial length L	180[mm]	160[mm]
Rim side blade inclination α	40[°]	40[°]
Hub side blade inclination β	40[°]	40[°]
Impeller outlet edge angle θ	15[°]	15[°]

An unstructured tetrahedral mesh is used through the domain. In the single-rotary model, the total nodes of channel 132387 and the total number of grids is 775472. In the double-rotary model, the total nodes of channel 124611 and the total number of grids is 712432. A grid independence check is performed. The calculating grids are as shown in figure 1.

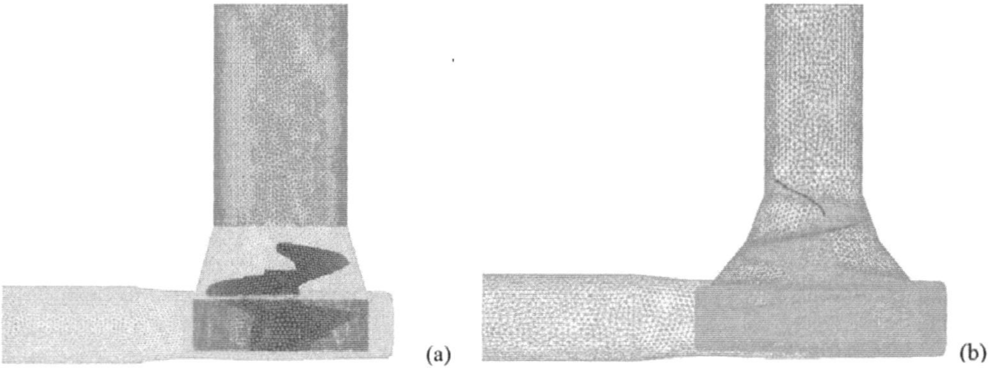

Figure 1. Calculating grid of (a) pump with single-rotary impeller and (b) pump with double-rotary impeller.

Results and Discussion

The streamlines of two different pumps with single-rotary impeller and double-rotary impeller are shown in figure 2. The results show that the flow field in the chamber of both pumps is rotating around the impeller shaft, and no obvious vortex is observed. The screw-type centrifugal pump features both a volumetric pump and a centrifugal pump, and the impeller has a large wrap angle and axial length, so the fluids flow at a fairly gentle manner in the chamber. As the fluids flow around the shaft in the chamber, it becomes a vertical flow in the outlet channel. A vortex occurs in the outlet channel of the pump with the single-rotary impeller, which is not observed in the channel of the double-rotary impeller. Compared to the double-rotary impeller, the flow channel of the single-rotary impeller is relatively short and the rectifying ability to the liquid is relatively weak. Velocity in the chamber is larger than in the inlet and outlet with both pumps. This difference is smaller in the single-rotary impeller and more pronounced in the double-rotary impeller. To explain this phenomenon, the flow rate distribution of a particular horizontal plane in the pump chamber is taken.

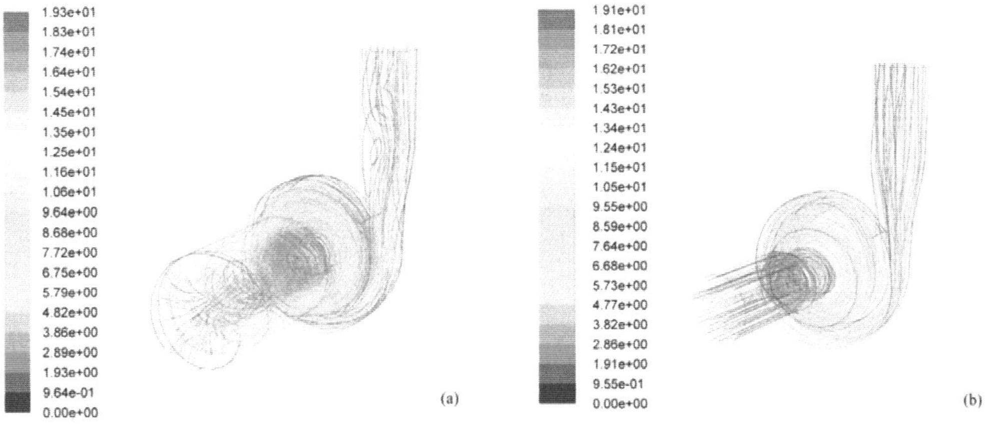

Figure 2. Streamlines of (a) pump with single-rotary impeller and (b)pump with double-rotary impeller.

The flow rate distributions corresponding to a horizontal plane in the pumps with single-rotary impeller and double-rotary impeller are shown in figure 3. As can be seen, the maximum velocities of the flow field in the chamber are 18.7 m/s and 19.1 m/s of pumps with single-rotary impeller and double-rotary impeller. There is no significant between the two data. But as shown in figure 3(a), it indicates that the maximum velocity only appears at the edge of the impeller, and the velocity is much smaller in the other part of the chamber. There is a similar phenomenon in the pump with the double-rotary impeller, although the difference of velocities between the edge and chamber is relatively small. The mean velocity in the chamber of the pump with single-rotary impeller is 4.51 m/s, as well as the mean velocity in the chamber of the pump with double-rotary impeller is 9.76 m/s. Larger flow velocity means greater pumping capacity, which is more beneficial in practical applications.

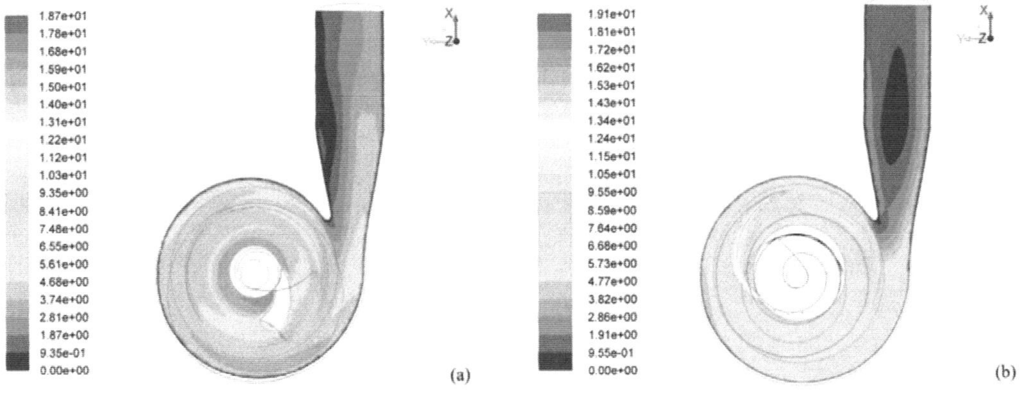

Figure 3. Velocity distribution in a horizontal plane of (a) pump with single-rotary impeller and (b)pump with double-rotary impeller.

Pressure distribution of pump with single-rotary impeller and double-rotary impeller are shown in figure 4. The maximum pressure of single-rotary type is 9.94*10^4 Pa, and the maximum pressure of double-rotary type is 3.90*10^4 Pa, which is much smaller than the single-rotary type. Higher pressure represents a stronger pumping capacity, which is more conducive to the delivery of complex and viscous media. In contrast, higher pressure makes the transported material more susceptible to damage. As mentioned earlier, the velocity distribution inside the double-rotary typer is more uniform than that of the single-rotary. Similarly, the pressure distribution inside the pump with the double-rotary impeller is more uniform than that of the single-rotary. The results may be due to that the flow channel of double-rotary type is longer than that of single-rotary type, therefore, the flow channel has a stronger binding force to the fluid.

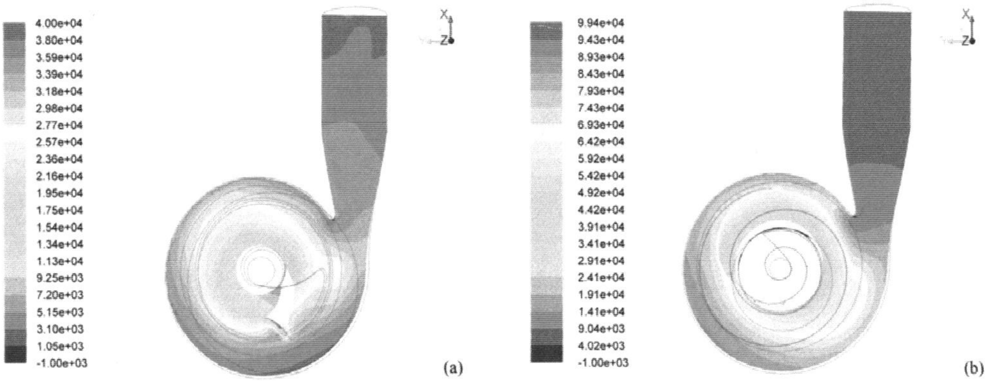

Figure 4. Pressure distribution in a horizontal plane of (a) pump with single-rotary impeller and (b)pump with double-rotary impeller.

Conclusion

In this study, the internal flow fields of screw-type centrifugal pumps with single-rotary and double-rotary impellers were simulated by CFD method. Streamlines, velocity, and pressure distributions were figured and discussed. It indicates that the velocities and pressure of the pump with the single-rotary impeller are both smaller than that with the double-rotary impeller. A larger flow velocity means greater pumping capacity, however, a higher pressure may lead to damages on the pumping materials. Both pumps have a relatively stable internal flow fields and are suitable for the transport of living things.

References

[1] Tatebayashi Y, Tanaka K and Kobayashi T 2005 *J. Turbom.* Pump Performance Improvement by Restraining Back Flow in Screw-Type Centrifugal Pump, **127(4)** 755-62.

[2] Han W, Ma W, Li R and Li Q 2012 *Proced. Eng.* The Numerical Analysis of Radial Thrust and Axial Thrust in the Screw Centrifugal Pump, **31** 176-81.

[3] Cheng X and Li R 2012 *Proced. Eng.* Parameter equation study for screw centrifugal pump, **31** 914-21.

[4] Kim D J, Suh S H and Sung S K 1998 *The KSFM J. of Fluid Machin.* Effects of the Impeller Shapes on the Non-Clogging and the Screw-type Centrifugal Pump Performances, **1** 81-9.

[5] Hui Q, Li R, Su Q, Han W and Wang P 2015 *Adv. Mech. Eng.* Analysis on Energy Conversion of Screw Centrifugal Pump in Impeller Domain Based on Profile Lines, **5** 512-23.

[6] Zubanov V, Volkov A, Matveev V, Popov G and Baturin O 2018 *Proceed. Asme Turbo Expo: Turbomachin. Technic. Conf. Expos.* Optimization of Fuel Two-Stage Screw Centrifugal Pump of Rocket Powerful Turbopump Unit, **2B** V02BT44A020.

[7] Zubanov V M, Shabliy L S and Krivcov A V 2014 *PROCEEDINGS OF THE ASME GAS TURBINE INDIA CONFERENCE* CFD-Modeling of Powerful Screw Centrifugal Kerosene Pump, **1** V001T01A002.

Acknowledgment
This work was financially supported by the National Key R&D Program of China (2016YFC1402306), Natural Science Foundation of Shandong (ZR2017MEE046), Shandong Key Research and Development Program (2016JMRH0541), Shandong Key Research and Development Program (2018GGX103018), Qingdao Basis Research of Application Program (16-5-1-101-jch) and Qingdao Basis Research of Application Program (17-1-1-97-jch).

The Numerical Analysis of Hydrodynamic Characteristics of Ducted Propeller by using SST *k*-*ω* Model

Wang Yue[1, 2, 3], Ren Wanlong[1, 2, 3], Liu Gang[1, 2, 3], Zhang Yuanming[1, 2, 3] and Hao Zongrui[1, 2, 3]

[1]Institute of Oceanographic Instrumentation, Qilu University of Technology (Shandong Academy of Sciences), Qingdao, 266100, China
[2]Shandong Provincial Key Laboratory of Marine monitoring instrument equipment technology, Qingdao, 266100, China
[3]National Engineering and Technological Research Center of Marine Monitoring Equipment, Qingdao, 266100, China

wangwill@vip163.com, haozr001@sina.com

Abstract. Hydrodynamic characteristics of the four-blade ducted propeller are numerically simulated by CFD method. The hydrodynamic performance of the propeller is analyzed by the SST *k*-*ω* Model, including the external characteristic parameters such as thrust and torque. The forward velocity at the propeller surface of the ducted propeller, the distribution of the induced velocity along the propeller surface, and the relationship between the propeller thrust and the surrounding velocity field are analyzed. The pressure distribution on the inner wall of the conduct is complicated. The pressure in the front half of the inner wall is small and large in the rear half. Axial, radial and tangential velocity distributions of sections at different distances from the paddle surface are calculated. The velocities on the dorsal side of the propellers are significantly higher than that on the foliar side and lower near the conduit.

1. Introduction

The ducted propeller is a commonly used propeller in underwater submersible systems. The trajectory control and navigational attitude adjustment of underwater submersible are generally accomplished by application of ducted propellers [1, 2]. In the case of large thrust and low speed, the propeller efficiency of the ducted propeller is significantly higher than that of the ordinary propeller. Therefore, for low-speed, high-thrust thrusters, ducted propellers are applicable [3]. The ducted propeller can also function as a rectification in use, reducing vibration and improving the stability of the ship sailing in the wind and waves. Also, ducted propellers are used to save energy [4].

As the application of ducted propellers is continually expanding, the computational fluid dynamics (CFD) method is introduced into the research of ducted propellers [5-7]. A lot of research and discussion have been carried out. The focus of these researches is the relationship between the flow field and thrust. Most studies use the slip grid to simulate the relative motion of the blades and the pipe in the water. By setting different boundary conditions to simulate the effects of different speeds and different flows on the thrust characteristics of the propeller. Many different models are used to do the calculations, such as standard *k*-*ε*, RNG *k*-*ε*, SST *k*-*ω*, etc. The pressure distribution, velocity distribution, and vortex shedding, cavitation are investigated. These studies are mainly on the characteristics of the propeller and on the surface of the propeller, however, there are few works studies the flow field at the specific plane behind the propeller disk. In this paper, the three-dimensional flow fields of the ducted propeller disk are simulated. Different planes behind the propeller disk are used to study the flow characteristics of the ducted propeller. The mechanism of complex flow region formation is discussed.

2. Computational Methods

2.1. Control Equations

2.1.1. Governing Equation of Fluid Flow. The laws of mass and momentum are applied for compressive laminar and turbulence flow. All control equations are represented by a conservation law. The coordinate system used to describe the problem is the Cartesian coordinate system in the Euler description. When the flowing fluid is defined as a continuum, the governing equations used to solve are as follows.

Continuity equation:

$$\frac{\partial \rho}{\partial t}+\frac{\partial}{\partial x_j}(\rho u_j)=0 \tag{1}$$

u is the instantaneous velocity in the j direction and ρ is the density of the fluid.

Momentum equation:

$$\frac{\partial}{\partial t}(\rho u_j)+\frac{\partial}{x_j}(\rho u_i u_j)=-\frac{\partial P}{\partial x_i}+\frac{\partial \tau_{ij}}{\partial x_j}+\rho f_i \tag{2}$$

P is the static pressure, τ_{ij} is the viscous stress tensor and f_i is the volume force. τ_{ij} can be expressed as:

$$\tau_{ij}=\left(\frac{\partial u_i}{\partial x_j}+\frac{\partial u_j}{\partial x_i}\right)-\frac{2}{3}\mu\left(\frac{\partial u_m}{\partial x_m}\right)\delta_{ij} \tag{3}$$

μ is the hydrodynamic viscosity, δ is the Kronecker function.

Bring (2) into (1) to get the Navier-Stokes equation:

$$\frac{\partial}{\partial t}(\rho u_j)+\frac{\partial}{\partial x_j}(\rho u_i u_j)=-\frac{\partial P}{\partial x_j}+\frac{\partial}{\partial x_j}\left(\mu\left(\frac{\partial u_i}{\partial x_j}+\frac{\partial u_j}{\partial x_i}\right)-\frac{2}{3}\mu\left(\frac{\partial u_m}{\partial x_m}\right)\delta_{ij}\right)+\rho f_i \tag{4}$$

2.1.2. SST k-ω Model. SST k-ω Model was proposed in 1993 by F.R. Menter [8]. As the Wilcox k-ω model is accurate in the near wall region and the k-ε model is suitable for freestream, this model combines these advantages of the Wilcox k-ω model and the k-ε model. The k-ε model is transformed into a k-ω model. The two models are weighted averaged by function F_1 and then added to obtain the BSL k-ω model. Based on the BSL k-ω model, the SST k-ω model modified the definition of the eddy-viscosity for adverse pressure gradient boundary-layer flows.

Original k-ω model and transformed k-ε model are weighted and averaged to obtain the following model:

$$\frac{\partial}{\partial t}(\rho k)=\tau_{ij}\frac{\partial u_i}{\partial x_j}-\beta^*\rho\omega k+\frac{\partial}{\partial x}\left[(\mu+\sigma_{k1}\mu_t)\frac{\partial k}{\partial x_j}\right] \tag{5}$$

$$\frac{\partial}{\partial t}(\rho\omega) = \frac{\gamma}{v_t}\tau_{ij}\frac{\partial u_i}{\partial x_j} - \beta\rho\omega^2 + \frac{\partial}{\partial x}\left[(\mu + \sigma_\omega\mu_t)\frac{\partial\omega}{\partial x_j}\right] + 2\rho(1 - F_1)\sigma_{\omega 2}\frac{1}{\omega}\frac{\partial k}{\partial x_j}\frac{\partial\omega}{\partial x_j} \tag{6}$$

Where,

$\mu_t = \frac{\rho a_1 k}{\max(a_{1\omega}, \Omega F_2)}$, $v_t = \frac{\mu_t}{\rho} = \frac{a_1 k}{\max(a_{1\omega}, \Omega F_2)}$, $F_2 = \text{tangarg}_2^2$, $arg_2 = \max\left(\frac{2\sqrt{k}}{\beta^*\omega d}, \frac{500v}{d_2\omega}\right)$, $F_1 = \text{tanharg}_1^4$, $arg_1 = min\left[max\left(\frac{\sqrt{k}}{\beta^*\omega d}, \frac{500v}{d_2\omega}\right), \frac{4\rho\sigma_{\omega 2}k}{CD_{k\omega}d^2}\right]$, $CD_{k\omega} = \max\left(2\frac{\rho\sigma_{\omega 2}}{\omega}\cdot\frac{\partial k}{\partial x_j}\cdot\frac{\partial\omega}{\partial x_j}, 10^{-20}\right)$

β^*, β, γ, σ_k, $\sigma_{\omega 1}$, $\sigma_{\omega 2}$, and a_1 are empirical constants. For any constant ϕ, there is

$$\phi = F_1\phi_1 + (1 - F_1)\phi_2 \tag{7}$$

The subscript 1 indicates the constant value in the inner layer, and the subscript 2 indicates the constant value in the outer layer.

2.2. Geometric Model Parameters and Meshing

The parameters of the simulated ducted propeller are listed as blow.

Table 1. Parameters of the simulated ducted propeller.

Propeller Diameter	Propeller solidity ratio	Number of blades
$R_p = 120$[mm]	$\Phi = 0.55$	$N = 4$
Inclination	Inlet diameter	Outlet diameter
$\alpha = 0$[°]	$R_i = 150$[mm]	$R_o = 150$[mm]

There are three methods for solving the problem of rotating motion commonly: dynamic mesh, slip mesh, and motion reference system settings. In this study, the method of motion reference system settings is chosen.

An unstructured tetrahedral mesh is used through the domain. The total nodes of channel 173413 and the total number of grids is 972572. The calculating grids are as shown in figure 1.

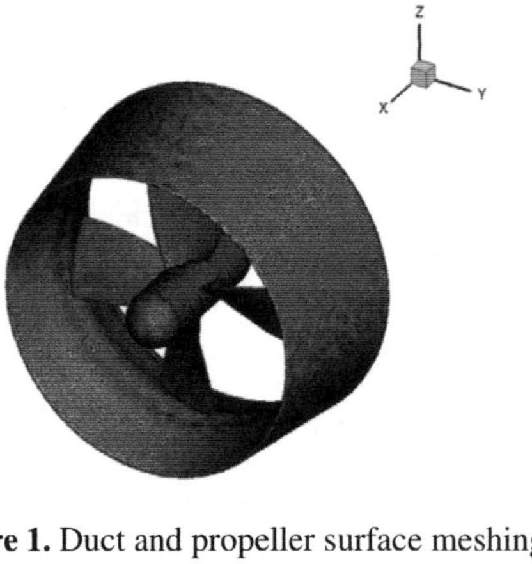

Figure 1. Duct and propeller surface meshing.

IOP Conf. Series: Materials Science and Engineering **649** (2019) 012025 doi:10.1088/1757-899X/649/1/012025

3. Results and Discussion

The pressure distribution of the duct at a speed coefficient J = 0.57 is shown in figure 2. It can be seen that the pressure distribution on the outer wall of the duct is relatively uniform. The axial force generated by the outer wall of the duct, i.e., the thrust, is substantially negligible. The pressure distribution on the inner wall of the duct is complicated. The pressure is small in the front half of the inner wall and large in the second half. There is a certain low-pressure area. Combined with the position and pressure distribution of the propeller, it is found that the low-pressure area of the inner wall of the duct is related to the position of the propeller. This is the location and source of the duct tip vortex.

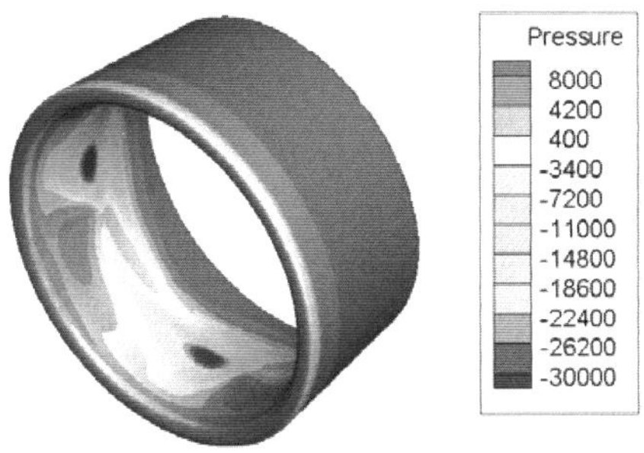

Figure 2. The pressure distribution of the duct at a speed coefficient J = 0.57.

The low-pressure zone is followed by the appearance of a high-pressure zone. This indicates that the propeller load has a large pressure influence on the duct, and the gap flow pressure fluctuation generated by the tip portion is complicated. Complex turbulence creates complex stress changes on the duct.

In order to analyze the internal flow field of the ducted propeller, the axial, radial and tangential velocity distributions at the different distances (x/R_p) from the paddle surface in the direction of the propeller axis are selected in the case of the inlet coefficient J = 0.5. The results at x/R_p = 0, 0.2 are shown in figure 3, figure 4, and figure 5.

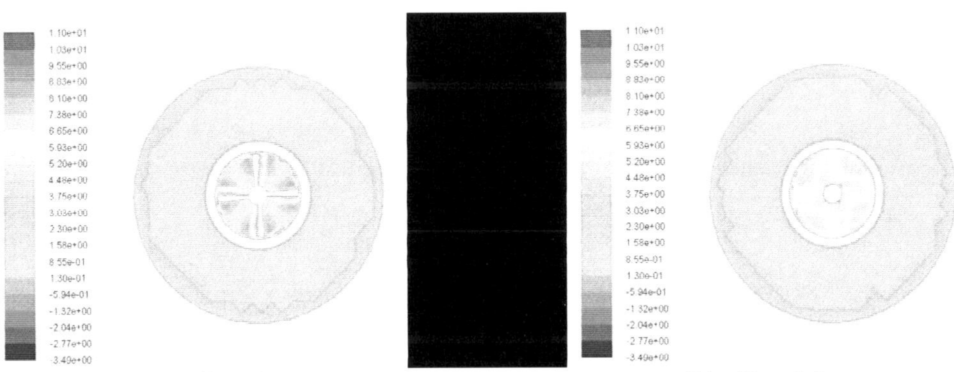

(a) $x/R_p = 0$ (b) $x/R_p = 0.2$

Figure 3. distribution of axial velocity.

Figure 3 shows the axial velocity distribution at the x/R_p = 0, 0.2 positions after the paddle surface. The axial velocity on the dorsal side of the propeller is significantly higher than that on the foliar side, and the velocity near the duct is low, in accordance with the Bernoulli equation analysis. The speed difference suggests that the foliar side is the pressure side and the back side is the suction side. The velocity gradient is larger at x/R_p = 0 and smaller at x/R_p = 0.2. This indicates that the viscous action between the fluids tends to converge the flow field.

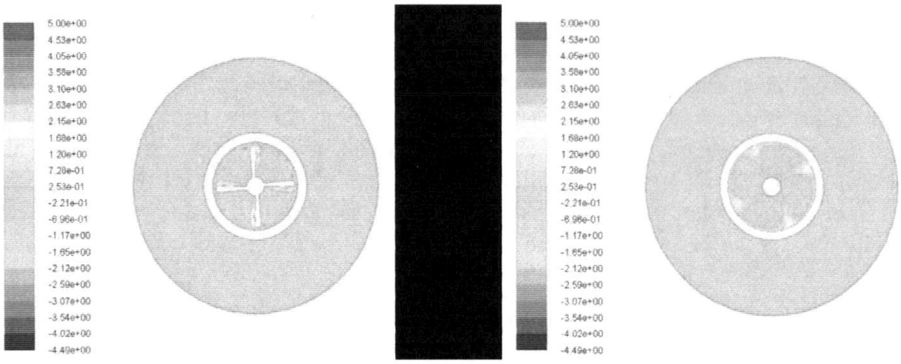

(a) $x/R_p = 0$ (b) $x/R_p = 0.2$

Figure 4. distribution of radial velocity.

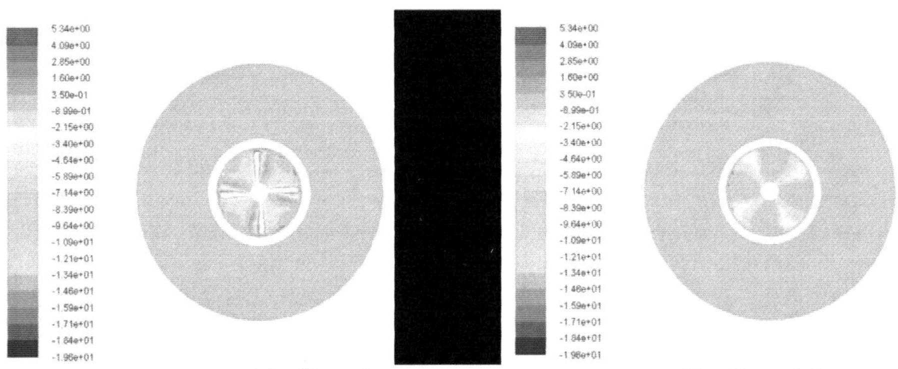

(a) $x/R_p = 0$ (b) $x/R_p = 0.2$

Figure 5. distribution of tangential velocity.

Figure 4 shows the radial velocity distribution at the $x/R_p = 0$, 0.2 positions after the paddle surface. The radial velocity of the flow field inside the duct is relatively small, and the outer zone is almost 0. Near the inner wall of the duct, a region with a small radial velocity is observed and mainly concentrated near the propeller tip. This is due to the fact that during the rotation of the propeller, the resulting tip vortex forms a complex flow area. The tangential velocity distribution is shown in figure 5, which is similar to the radial velocity distribution.

4. Conclusion

In this study, the flow fields of the ducted propeller were simulated by CFD method. Axial velocity, radial velocity, and tangential velocity distributions were figured and discussed. It indicates that The axial velocity on the dorsal side of the propeller is significantly higher than that on the foliar side, and the velocity near the duct is low, in accordance with the Bernoulli equation analysis. Near the inner wall of the duct, a region with a small radial velocity is observed, which is due to the complex flow area formed by the rotation of the propeller.

References

[1] AVILA J P J and ADAMOWSKI J C 2011 *Ocean Eng.* Experimental evaluation of the hydrodynamic coefficients of a ROV through Morison's Equation, **38(17-18)** 2162-70.

[2] JAULIN L 2009 *Automatica* Robust set-membership state estimation: application to underwater robotics, **45(1)** 202-6.

[3] Gong J, Guo C, Zhao D, Wu T and Song K 2018 *Ocean Eng.* A comparative DES study of wake vortex evolution for ducted and non-ducted propellers, **160** 78-93.

[4] Mysa R C, Tuyen L Q, Ma A and Nguyen V T 2018 *Proceedings of The ASME 37th Int. Conf. Ocean, Offshore Arctic Eng.* Numerical investigation of energy saving device effects on ships with propeller-hull interactions, **7B** V07BT06A044.

[5] Gaggero S, Villa D, Tani G, Viviani M and Bertetta D 2017 *Ocean Eng.* Design of ducted propeller nozzles through a RANSE-based optimization approach, **145** 444-63.

[6] Vasudev K L, Sharma R and Bhattacharyya S K 2018 *Ships Offsho. Struct.* Shape optimisation of an AUV with ducted propeller using GA integrated with CFD, **13(2)** 194-207.

[7] Andersson J, Eslamdoost A, Vikstrom M and Bensow R E 2018 *Ocean Eng.* Energy balance analysis of model-scale vessel with open and ducted propeller configuration, **167** 369-79.

[8] Menter F R 1993 *23rd Fluid Dynamics, Plasmadynamics, Lasers Conf. Fluid Dynamics and Co-located Conf.* Zonal Two Equation k-ω Turbulence Models for Aerodynamic Flows, **6** AIAA-93-2906.

Acknowledgment

This work was financially supported by the National Key R&D Program of China (2016YFC1402306), Natural Science Foundation of Shandong (ZR2017MEE046), Shandong Key Research and Development Program (2016JMRH0541), Shandong Key Research and Development Program (2018GGX103018), Qingdao Basis Research of Application Program (16-5-1-101-jch) and Qingdao Basis Research of Application Program (17-1-1-97-jch).

Numerical Analysis of the Influence of Superconductor and Magnet Material Parameters on the Dynamic Stability of Maglev Systems

Chen-Guang Huang[1] and Bin Xu[1]

[1]127 West Youyi Road, Beilin District, Xi'an Shaanxi, 710072, P.R. China

huangcg@nwpu.edu.cn, xubin@nwpu.edu.cn

Abstract. In previous research, a two-dimensional numerical model based on Newton's second law and Maxwell's equations was developed to analyze the dynamic behavior of superconducting maglev systems. Studies showed that vibration and drift phenomena will occur in the vertical and lateral directions triggered by external disturbances, resulting in a change in the levitation point and even instability. In this paper, a numerical analysis is further carried out to alleviate these situations. The dependence of the drifts of vertical and lateral vibration centers of the levitated body upon system material parameters, such as the magnetization of permanent magnets and the critical current density of superconductor, is studied in terms of a power law model. In addition, the influence of the configuration of permanent magnets on the dynamic stability is analyzed.

1. Introduction

The magnetic levitation realized by superconductors (SCs) and permanent magnets (PMs) has resulted in a wide range of applications including industrial levitation devices with the advantages of self-stability with no active control and no energy consumption. Among these devices, there is a great interest in maglev systems where bulk high-temperature SCs attain the levitation of vehicles above PM guideways with longitudinal geometry, such as linear magnetic bearings, launching and transportation systems [1, 2], such that, due to the huge potential for application, improving their levitation performance becomes particularly important. Some experiments have reported that the levitation force exhibits time-dependent relaxation, and the downward drift of vibration center of levitated body will occur under external disturbances. The thermally activated flux motion, an intrinsic feature of the SC, is regarded as the primary cause of these nonlinear behaviors. Meanwhile, numerous numerical studies have tried to understand the dynamic levitation by considering the flux motion by means of different numerical methods such as the T-method, control volume method, and finite element method [1-3]. Many of these studies mainly focus on the disturbance-related levitation of the maglev by assuming the SC moving along some virtual displacement. Such displacement is not necessarily equal to the one followed by the system after a perturbation of equilibrium. However, from the perspective of dynamics, the motion trajectory of the SC subjected to external disturbances should generally be expected to be a complex spatial curve under the combined action of gravity and magnetic forces. Based on Newton's second law and the A-V formulation, recently, we have developed a two-dimensional numerical model to analyze the dynamics of maglevs and found that vibration and drift phenomena will occur in the lateral and vertical directions triggered by external disturbances. However, there is not a complete numerical study that analyzes how the dynamic stability of these maglevs depends on the system material parameters or the guideway configuration. In this paper, we use the same model to further study the dynamic behavior of several typical maglev systems. The aim of this paper is to reveal the influence of the SC and PM properties, such as magnetization, critical current density, and configuration, on the dynamic stability and establish some general guidelines on resisting external disturbances.

2. Model Description

We consider maglev systems consisting of a bulk high-temperature SC and a guideway formed by different sets of PMs, as shown in figure 1. The PM configurations are distinguished as follows: PM2 and PM2L stand for two magnets with vertical and antiparallel horizontal magnetization, respectively; PM3L denotes three magnets with vertical magnetization in the central magnet and antiparallel horizontal magnetization in the lateral ones; and PM3N indicates three magnets with antiparallel vertical magnetization. In these systems, the SC and PM are infinitely long in the z direction and have rectangular cross-sections of a_{SC}*b_{SC} and a_{PM}*b_{PM} in the x and y directions, respectively. The SC has a critical current density J_c and the PMs have a magnetization M with a horizontal interval a.

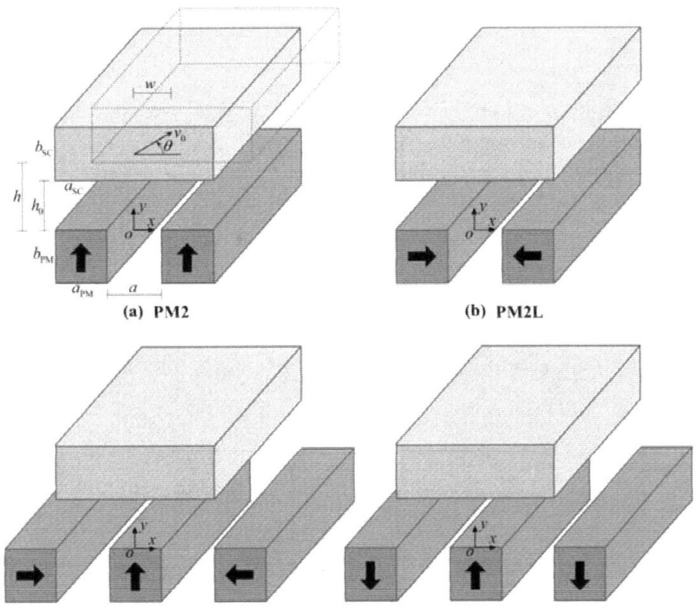

Figure 1. Sketch of the studied maglev systems.

We assume that the SC is cooled at very large height above the PMs, and then it is forced downwards with a small velocity 10^{-4} m/s to a static equilibrium height h_0, i.e., the working height. Once reaching such a height, the dynamic motion of the SC is triggered by an external disturbance. Using Newton's second law of motion, the equations of lateral and vertical movements of the SC can be expressed as

$$mw''-F_x=0 \quad (1)$$

$$mh''+mg-F_y=0 \quad (2)$$

and the initial conditions are taken into account by

$$t=0:\ w=0,\ h=h_0,\ w'=v_x=v_0\cos(\theta),\ h'=v_y=v_0\sin(\theta) \quad (3)$$

where m is the mass per unit length of the SC, v_0 is the disturbance-induced velocity with a tilt angle θ, g is the gravitational acceleration, and F_x and F_y are the horizontal (guidance) and vertical (levitation) magnetic forces per unit length on the SC, respectively, traditionally calculated by Lorentz's equation (3).

In general, the Maxwell's equations are used to describe macroscopic electromagnetic phenomena. By adopting the quasi-static approximation of the Maxwell's equations, the following formulation is obtained.

$$\mathbf{E}=-\mathbf{A}'-\nabla V \quad (4)$$

$$\nabla^2\mathbf{A}=-\mu_0\mathbf{J} \quad (5)$$

where μ_0, V , **E**, **A** and **J** are the permeability, electric potential, electric field, vector potential and current density, respectively. For the SC, it has a highly nonlinear current-voltage characteristic and the material constitutive relation between **E** and **J** can be expressed with a power-law as follows.

$$\mathbf{E}=E_c(\mathbf{J}/J_c)^n \quad (6)$$

where E_c is the critical electric field and n is the creep exponent. It is obvious that once the levitation position is changed, the SC will undergo a magnetic field variation and then its induced currents redistribute. Thus, the magnetic forces also adjust themselves to the change of the magnetic field and induced currents. On the other hand, the adjustment of the magnetic forces causes a change in the levitation position. That is to say, the dynamics of the maglev system is coupled with superconducting phenomena. Recently, we have proposed an iterative method to solve similar coupling problems. More details on the numerical procedure to deal with equations (1)-(6), including the theoretical validation, can be found in Refs. [4] and [5].

Table 1. Parameter values used in the numerical simulation [3-5].

Creep exponent	$n=21$
Critical electric field	$E_c=10^{-4}$ V/m
Disturbance-induced velocity	$v_0=0.01$ m/s
Tilt angle of disturbance-induced velocity	$\theta=45^0$
Characteristic critical current density	$J_0=3.7*10^6$ A/m^2
Characteristic magnetization	$M_0=7.95*10^5$ A/m
Geometry of superconductor	$a_{SC}/3=b_{SC}=a=0.05$ m
Geometry of permanent magnets	$a_{PM}=b_{PM}=a=0.05$ m
Working height of superconductor	$h_0=0.5a=0.025$ m

3. Results and Discussion

Figure 2 displays the calculated field lines and current profiles for PM2L with different Jc and M at the working height. One sees that when the SC is moved vertically downward, flux vortices created by the PMs will mainly penetrate into the bottom region of the SC. Such penetrated region will be reduced with increasing Jc, which leads to an increase in the induced currents and further in the levitation force (see figures 2(a), (b) and (c)). Besides, an increase in M can attain an intensification of the magnetic field and flux gradient above the guideway, which results in larger field-penetrated region and induced currents and thereby bigger levitation force (see figures 2(a), (d) and (e)).

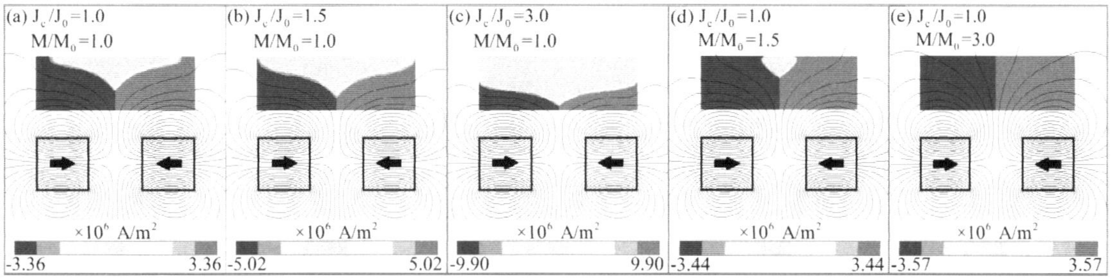

Figure 2. Field lines and current profiles for PM2L with different J_c and M at the working height.

Figure 3 shows the dynamic response in PM2L with different J_c and M under the external disturbance. In figure 3, one sees that after being disturbed, the SC will be simultaneously vibrated in the x and y directions, and the vertical and lateral vibration centers gradually drift along their vibration directions. Since flux vortices move in and out during vibration, resulting in energy dissipation, the vibration amplitudes in both x and y directions decrease with time little by little. The results of figures 3(a) and (b) show that with the increase in J_c, the drift of vertical vibration center is monotonously decreased. This is due to that increasing J_c can lead to an increase in the values of the induced currents in the field-penetrated region. From figures 3(c) and (d), one sees that for the case of larger M, the PMs have a stronger magnetic field and a higher flux gradient at their upper surfaces, tending to produce larger magnetic forces. In the y direction, due to the hysteresis effect, the levitation force will rapidly decay with time. Thus, the drift of vertical vibration center shows a monotonic increase with increasing M.

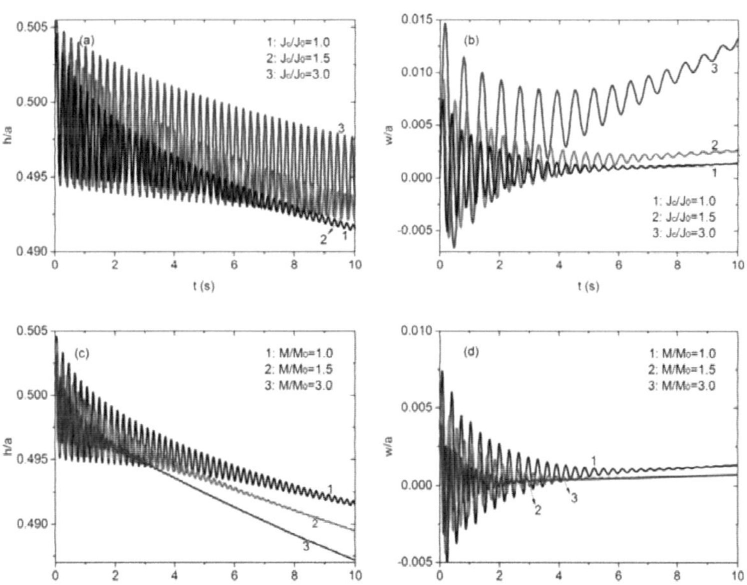

Figure 3. Time responses of levitation height and lateral displacement for PM2L with different J_c and M.

Figure 4 shows the dependence of the SC's dynamic response on the guideway configuration at $M=M_0$ and $J_c=J_0$. It is seen that for configurations with two PMs, PM2L is a good choice for application as it obtains a large levitation force and a small drift for vertical vibration center, although it exhibits a slightly larger drift for lateral vibration center relative to PM2. However, for a guideway composed of three PMs, the levitation force provided by PM3L is much larger than that provided by PM3N. However, the capacity of resisting external disturbance for the latter is superior to that for the former. Thus, one needs to make a tradeoff between the levitation force and dynamic stability when selecting such two configurations.

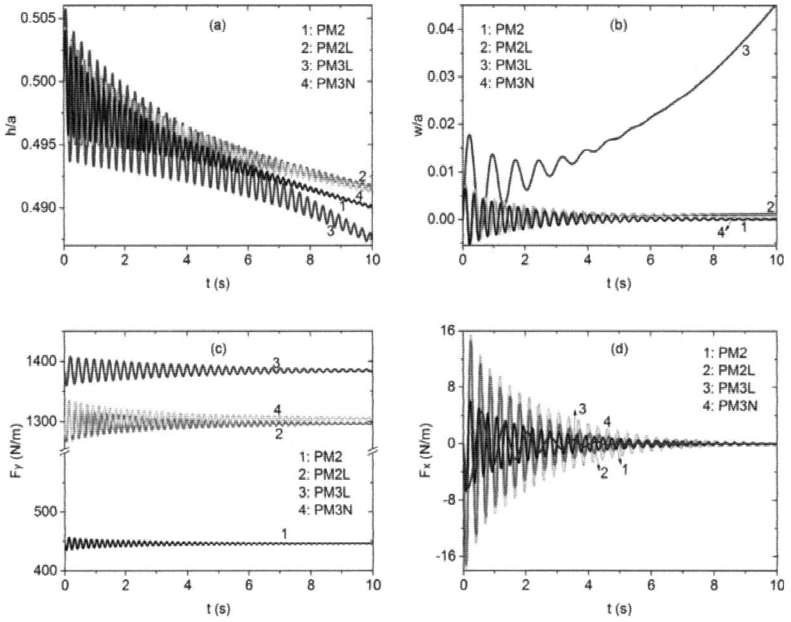

Figure 4. Time responses of levitation height, lateral displacement, levitation and guidance forces for the SC in the four studied systems with $J_c=J_0$ and $M=M_0$.

Summary

We present a systematic study of the dependence of the vibration and drift phenomena triggered by external disturbances in typical maglev systems composed of a bulk SC and a guideway formed by different sets of PMs upon the system material parameters. Results show that although the levitation force can be increased by modifying system material parameters, this may entail a reduction in the stability. For example, increasing the SC's critical current density leads to a decrease in the drift of vertical vibration center but an increase in that of the lateral vibration center. Similarly, increasing the PMs' magnetization causes the drift of vibration center to decrease in the lateral direction but that to increase in the vertical direction. Moreover, we find that for a guideway formed by two or three PMs, the configuration PM3N appears to be the best choice in a maglev application since it provides a large levitation force with good dynamic stability.

References

[1] Wang J, Wang S, Zeng Y, et al. 2002 *Physica C Supercond.* The first man-loading high temperature superconducting Maglev test vehicle in the world, **378-381(Part1)** 809-14.
[2] Werfel F N, Floegel-Delor U, Rothfeld R, Riedel T, Goebel B, Wippich D and Schirrmeister P 2011 *Supercond. Sci. Technol.* Superconductor bearings, flywheels and transportation, **25(1)** 014007.
[3] Ma G T, Ye C Q, Liu K, Mei G M, Zhang H and Li X T 2016 *IEEE Trans. Appl. Supercond.* Geometrical effects on the levitation capability of multiseeded Y-Ba-Cu-O blocks, **26(4)** 3600205.
[4] Huang C G, Xue C, Yong H D and Zhou Y H 2017 *J. Appl. Phys.* Modeling dynamic behavior of superconducting maglev systems under external disturbances, **122(8)** 083904.
[5] Huang C G, Xu B and Zhou Y H 2019 *Supercond. Sci. Technol.* Dynamic simulations of actual superconducting maglev systems considering thermal and rotational effects, **32** 045002.

Numerical Simulation on Oil Spill at Different Positions on the Back Surface of a Blunt Body

Juan Xu, Zongrui Hao, Yue Wang, Jin Liu, Gang Liu and Yuanming Zhang

[1]Institute of Oceanographic Instrumentation, Qilu University of Technology (Shandong Academy of Sciences), Qingdao, 266100, China
[2]Shandong Provincial Key Laboratory of Marine monitoring instrument equipment technology, Qingdao, 266100, China
[3]National Engineering and Technological Research Center of Marine Monitoring Equipment, Qingdao, 266100, China

xujuan101066@163.com

Abstract. The movement of oil spilled from the back surface of an underwater blunt body is influenced by the flow around the blunt body, so the trajectory of the oil movement changes with the vortex behind the blunt body. A numerical modelling using VOF model is established to simulate the oil spill process. Three typical spilling points at the top, middle, and bottom of the back surface of the blunt body, respectively, are simulated at different leaking rates. Their vorticity fields and the oil trajectories are analyzed. The results show that the spill from the top position generates the smallest oil distribution on the sea surface, while that from the bottom position leads to the largest one.

1. Introduction

Offshore oil can be spilled into the ocean, river and other waters in case of accidents in the oil exploitation and transportation, which is harmful to the water environment. At present, there is no effective means to prevent the occurrence of oil spill [1]. Oil spilled from ships and offshore platforms have become an important issue affecting the marine environment. According to statistics, about 6 million tons of oil is released into the ocean every year over the world. From 1974 to 1979, there were more than 4,000 oil spill accidents in the world. And from 1967 to 1991, 54 major accidents with the spilled oil over 10,000 tons each were reported [2].

The movement of spilled oil is complicated, and is affected by a variety of environmental factors. Many models on this subject have been developed over the past 40 years. Shen and Yapa [3] established the oil film transport model, Ross, based on the surface oil film in the river. On this basis, Yang and Shen [4] et al established a two-layer mathematical model, Ross2, for oil transportation in rivers. Ross2 regards the suspended oil beads which is dispersed into the water as a suspended layer, and there is constant mass exchange between the surface layer and suspended layer. Li [5] described the movement of oil droplets based on the convection-diffusion equation, and a diffusion term caused by the rising oil droplets was added. In the mid-1980s, Johansen et al [6] developed the "oil particle" model. This model showed good performance in simulating oil leakage with pretty accurate reproduction of the oil diffusion and distribution. Papadimitrakis et al [7] used two-phase flow equation to simulate the dynamic process of oil spill, and the weathering process was also taken into account in the model. To

IOP Conf. Series: Materials Science and Engineering **649** (2019) 012027 doi:10.1088/1757-899X/649/1/012027

sum up, many works on the modelling of oil leakage has been reported, however, no researchers have been carried out on the oil spill from the back surface of a blunt body.

In this paper, a numerical simulation on oil spill at different locations, top, middle and bottom (e.g. h, h/2 and 0) on the back surface of a blunt body (oil tank) is presented. The influences of the flow around the blunt body and the oil spilling rate were investigated. This study will be beneficial to the treatment of oil spill accidents.

2. Numerical Theory

2.1. Governing Equations

For the free surface flow problem, the fluid is incompressible, the viscosity coefficient of the fluid is a constant, and the governing equation is a simplified N-S equation. The coordinate system used to describe the problem is the rectangular coordinate system described by Eulerian method. The horizontal direction is the X-axis, the right direction is positive, the vertical direction is the z-axis, and the upward direction is positive. u, w are the velocity components of x and y direction, ρ is the density of the fluid, p is the pressure of the fluid, and μ is the coefficient of kinematic viscosity of the fluid [8].

The continuity equation is as follows.

$$\frac{\partial u}{\partial x}+\frac{\partial w}{\partial z}=0 \tag{1}$$

The momentum equation is as follows.

$$\frac{\partial u}{\partial t}+u\frac{\partial u}{\partial x}+w\frac{\partial u}{\partial z}=-\frac{1}{\rho}\frac{\partial p}{\partial x}+v[\frac{\partial^2 u}{\partial x^2}+\frac{\partial^2 u}{\partial z^2}] \tag{2}$$

$$\frac{\partial w}{\partial t}+u\frac{\partial w}{\partial x}+w\frac{\partial w}{\partial z}=-\frac{1}{\rho}\frac{\partial p}{\partial x}+v[\frac{\partial^2 u}{\partial x^2}+\frac{\partial^2 u}{\partial z^2}] \tag{3}$$

2.2. Boundary Conditions

The normal velocity at the bottom of the water area is zero, i.e., the impenetrable condition is used, which is as follows.

$$\frac{\partial \emptyset}{\partial n}|_{z=-d}=0 \tag{4}$$

Here n is the outer normal direction of the boundary.

The kinematic boundary conditions of the free surface are as follows.

$$\frac{\partial \emptyset}{\partial z}|_{z=\eta}=\frac{\partial \eta}{\partial t}+\frac{\partial \eta}{\partial x}\frac{\partial \emptyset}{\partial x}|_{z=\eta}+\frac{\partial \eta}{\partial y}\frac{\partial \emptyset}{\partial y}|_{y=\eta} \tag{5}$$

$z=\eta(x,y,t)$ is the equation of the free surface.

The dynamic boundary condition of the free surface is as follows.

$$\frac{\partial \emptyset}{\partial n}|_{z=\eta}+g\eta+\frac{1}{2}(\nabla\emptyset\cdot\nabla\emptyset)\frac{\partial \eta}{\partial t}|_{z=\eta}+\frac{p}{\rho}=0 \tag{6}$$

Based on the above equations, a mathematical model can be established to simulate the movement of different sea water under different conditions.

2.3. The VOF Model

The VOF method is usually used to solve the water - gas interface. The water surface of the tank belongs to the gas-water two-phase interface, and the underwater oil spill belongs to the three-phase flow, i.e., oil, water and gas, in which adopting the VOF method is suitable [7].

The VOF model simulates two immiscible fluids by separately solving the momentum equation and processing the volume ratio of each fluid passing through the control volume. The equation is as follows.

$$\frac{\partial}{\partial t}(\rho v) + \nabla(\rho v v) = -\nabla p + \nabla \cdot [\mu(\nabla v + \nabla v^T)] + \rho g + F \tag{7}$$

2.4. Geometric Model

The submarine oil tank position is shown in figure 1, with the calculation domain length of 30 m, width of 12 meters, water depth of 10m, and the diameter of oil spilling hole of 0.05 m. The spilling hole is on the back surface of the pipeline or oil tank opposite to the flow direction. Three spilling positions, which are shown as Positions A, B and C in figure 1, are considered in this study. The height of the three positions are h, h/2 and 0, respectively (the lower boundary is 0, and the height of the blunt body is h). The oil spilling rate is set as 0.2m/s. Quadrilateral grid is used in the whole geometric region, and the grid number is 157681. The grid diagram is shown in figure 1.

Fluent software was used, and Segregated solver, Standard k-ε model and PISO algorithm were selected to solve the turbulence.

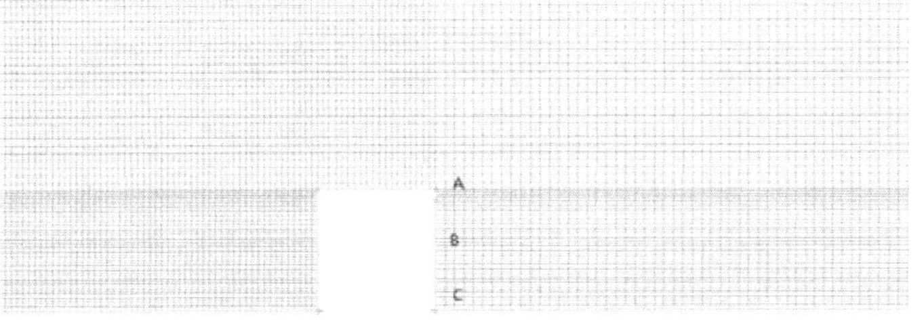

Figure 1. Geometric model and grid.

3. Results and Analysis

3.1. Trajectory of Spilled Oil

When oil spill occurs underwater in an accident, due to the unsteady flow separation around the blunt body (oil tank or pipes), vortexes are generated downstream. The vortex has an impact on the movement of spilled oil, and hence movement trajectory of the spilled oil changes accordingly [8]. Figure 2 shows the phase distributions at T=50s for the three cases with different oil spilling position when the oil spills at a rate of V=0.2m/s. It is apparent that the spilling position has an important influence on the distribution of oil spill. In the same period, the lower the spilling position is, the larger area the spilled oil spreads on the water surface.

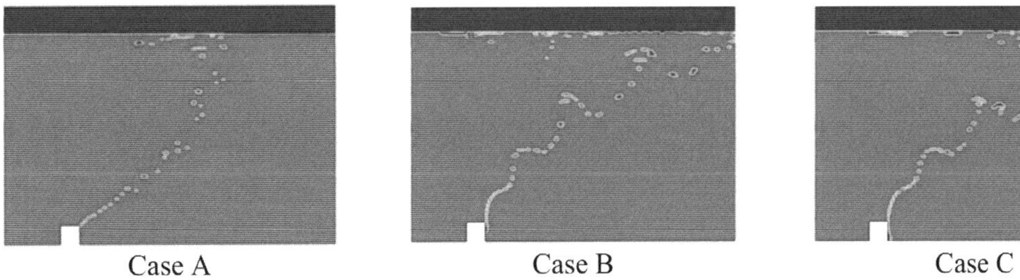

Case A Case B Case C

Figure 2. The movement trajectory of spilled oil at spilling rate $V = 0.2\text{m/s}$.

IOP Conf. Series: Materials Science and Engineering **649** (2019) 012027 doi:10.1088/1757-899X/649/1/012027

3.2. Vorticity Field

Vorticity in a two-dimensional flow field is as follows.

$$\Omega=\frac{\partial v}{\partial x}-\frac{\partial u}{\partial y} \tag{8}$$

For the three spilling positions A, B and C with the same oil spilling rate V =0.2m/s and the same sea water velocity V =0.2m/s. The vorticity field distributions for different spilling positions are shown in figure 3. Take the spilled oil as the dividing line, the right side is affected by the vortexes downstream the blunt body while the left side only is only slightly or not affected. For the Position A, due to the high spilling position, the spilled oil goes up straightforwardly under the effect of buoyancy and is almost not affected by the vortexes behind the blunt body during the rising period. The oil spilled from Positions B and C will go through the high-energy vortexes area downstream the blunt body during their rising. Therefore, after interacting with the vortexes, the spilled oil moves up along and closely to the wall surface. When leaving the wall surface, the oil oscillates downward under the alternating action of sea flow and vortex. After a while, horizontal and vertical displacement occurs, the oil rises in the twist. Compared with Positions A and B, Position C is greatly affected by vortexes and alternating vorticities appears not far from the blunt body, indicating secondary vortexes are generated. Since Position B is at half height of the blunt body, the spilled oil interacts with the vortex with different strength from Position C. Therefore, Position B shows different secondary vortex's torsion, as can be seen from figure 3.

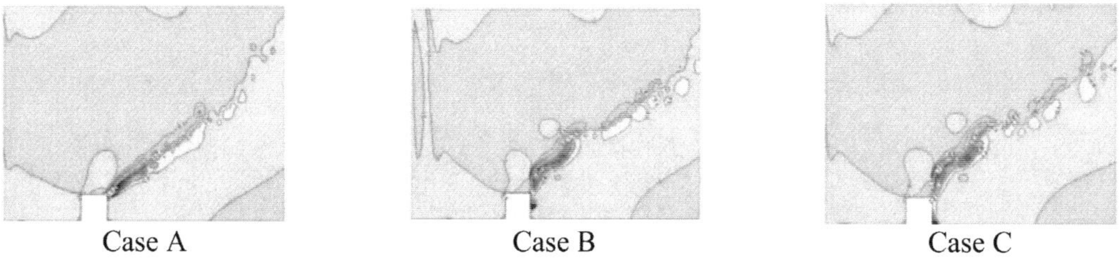

Figure 3. Vorticity fields for different oil spilling postions.

3.3. Velocity field

Figure 4 shows the velocity vectors of the oil spilling processes at three different positions. Despite of the different spilling position, the main flow velocity of the spilled oil close to the blunt body is determined by the spilling rate. The oil spilled from position A rises instantaneously and moves upward under the effect of the vortex behind the blunt body. After moving away from the vortex, the oil moves under the effect of buoyancy and sea water flow. In contrast, the oil spilled from Position C is greatly affected by the vortexes, as a result, it has greater upward velocity and will firstly rise up along the wall surface when it is spilled out. Then the oil deflects away from the blunt body under the effect of sea water flow after it reaches the top surface of the blunt body. Comparing with spilling position C, Position B is in the middle of the blunt body, so the upward velocity of the spilled oil along the wall surface is smaller, hence it deflects immediately after leaving the spilling port on the back surface of the blunt body.

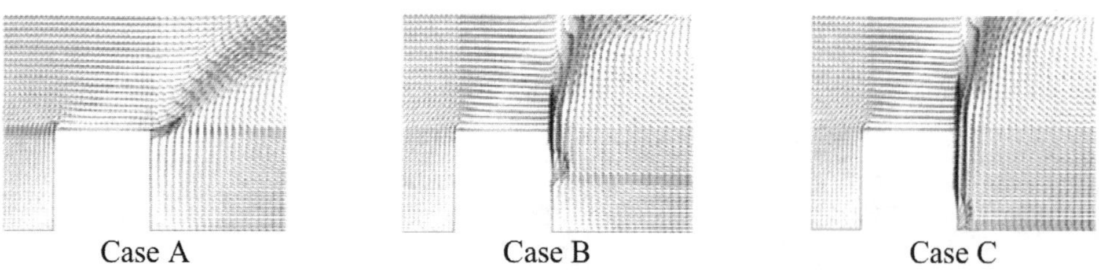

Figure 4. Vorticity fields for different oil spilling postions.

4. Conclusion

A numerical model including VOF model and transportation model was established to simulate the underwater oil spill. The oil spills from different postions on the back surface of a blunt body were simulated. Through the analysis on the moving trajectory of the spilled oil and the vorticity and velocity field of the calculation domain, the following conclusions were obtained.

1. It is feasible to use VOF model and transportation model to analyze the oil spill from the back of a blunt body. The results are intuitive, which can provide a basis for emergency management of oil spill underwater.

2. Depending on the spilling position, the vortex behind the blunt body has different influence on the oil spill. For the oil spilled from the bottom and middle positons, the oil rises close to and along the wall with almost no horizontal movement. Only after reaches the top surface of the blunt body, the oil moves horizontally with the sea water flow.

3.The lower the spilling position is, the more of the spilled oil is affected by the vortex behind the blunt body, which results in a longer vertical displacement time of the oil spill and a larger spread of spilled oil on the sea surface.

References

[1] Maowen C E and Jianping W E N 1993 *Beijing: China Sci. Technol. Press* Global environmental problems and countermeasures.

[2] Wei-zheng Q U and DENG S G 2001 *J. Natural Disasters* Disastrous ocean pollution of petroleum, **10(1)** 69-74.

[3] Shen H T and YaPa P D 1987 *Water Resource Res.* Asimulation model for oil slick transport lakes, **23(10)** 1949-57.

[4] Yang X Q, Shen H D and Wang D S 1996 *J. Hydraulic Eng.* A Two-layer mathematical model for oil slick transport in river, **(8)** 71-6.

[5] Li W 1999 Progress in water environmental science [M]. Wu Han: Wuhan university of water resources and hydropower press.

[6] ASCE Task Committee on modeling of oil spills of the water resources engineering division 1996 *J. Hydraulic Eng.* State-of-the-art review of modeling transport and fate of oil spills, **122(11)** 594-609.

[7] Papadimitrakis J, Psallaki M, Christolis M, et al. 2006 *Environ. Modelling & Software* Smulating the fate of an oil spill near coastal zones: The case of a spill (from a Power Plant) at the Greek Island of Lesvos, **21(2)** 170-7.

[8] ZHENG J Y, LIU Y L, XU P, et al. 2009 *J. Zhejiang University (Eng. Sci.)* Numerical simulation of obstacle influence on leakage and diffusion of hydrogen due to high-pressure storage tank failure, **22(3)** 265-70.

Acknowledgements

This work was financially supported by the National Key R&D Program of China (2016YFC1402306), Natural Science Foundation of Shandong (ZR2017MEE046), Shandong Key Research and Development Program (2016JMRH0541), Shandong Key Research and Development Program (2018GGX103018), Qingdao Basis Research of Application Program (16-5-1-101-jch) and Qingdao Basis Research of Application Program (17-1-1-97-jch).

A Study on Materials and Development of Wireless Mobile Communication

Byungseok Kang[1] and Kijong Yoo[2]

[1]Dept. of Electronics, Computing and Mathematics, University of Derby, Derby, DE22 1GB UK
[2]HyoMyeong High School, 51 Songtan-ro, Pyeongtaek-si, Gyeonggi-do, 17784, Republic of Korea

b.kang@derby.ac.uk, mathink@naver.com

Abstract. In modern society, the need for rapid information exchange is increased continuously. As a result, the telecom industry has grown rapidly to more than double size of the global automotive market and has a major impact on the global economy. With the rapid development of the wireless communication market, a new concept of semiconductor materials having better physical and electrical characteristics. In this paper, we study on relationship between material engineering and wireless communication system. We introduce a greedy heuristic algorithm which optimize the wireless network construction for mobile streaming systems.

1. Introduction

The development of the mobile communications semiconductor industry is based on new semiconductor materials and semiconductor implementation process technology. Currently, a power amplifier using gallium arsenide is embedded inside the mobile phone terminal [1]. Device studies using gallium nitride and silicon carbide are underway to realize power amplifiers for base stations that require greater power amplification than mobile handsets and mobile communications semiconductor components that operate normally under conditions of high temperature and high pressure.

Semiconductor materials play an important role in manufacturing wireless terminals. Production of all semiconductor components such as DRAM and power amplifiers begins with semiconductor single crystal growth. Currently the most widely used semiconductor material is silicon (Si). When silicon oxide (SiO2), which is also the main component of sand, is reacted with carbon at high temperature, polycrystalline silicon having 95-97% purity can be obtained. Further, after a finer refining process, polycrystalline silicon having a purity of 99.99% or more is obtained. Polycrystalline silicon is made up of numerous fine crystals. At the interface between the crystals, the atoms are not regularly arranged. This causes the movement of electrons in the semiconductor. Therefore, in order to be used as a semiconductor device material, a monocrystalline silicon having no interface is required.

The polycrystalline semiconductor is immersed in a crucible having a high melting point such as graphite or quartz, and then is heated to a temperature of more than the melting point of the semiconductor to form a liquid state [2]. When the seed crystal is immersed in liquid phase and then slowly pulled up, single crystals grow between the liquid phase and the seed crystal to form a single crystal semiconductor. When the cylindrical surface of the cylindrical single crystal thus formed is ground and made uniform, a flat surface for indicating the direction is formed, and then cut to a thickness of 0.5-0.7 mm using a diamond saw, a semiconductor wafer is produced.

However, such a semiconductor wafer is not directly used for device fabrication. Instead, it is used as a substrate to make a more precisely controlled semiconductor thin film. This is called epitaxy growth. There are physical and chemical methods for epitaxial growth. The physical method is to charge the semiconductor constituent material into an ultrahigh vacuum reactor of 1*10-10torr or less, and then heat the crucible containing the semiconductor material so that the vaporized semiconductor atoms reach the substrate to form crystals. The chemical method refers to the growth of a crystal thin film layer by causing a gas containing a semiconductor constituent substance to cause a chemical reaction at a high temperature wafer surface. The semiconductor thin film grown by the epitaxial method has perfect crystallinity.

There has been much work on material and wireless streaming systems in the literature [3-7]. Based on the overlay topology construction, these peer-to-peer streaming systems can be categorized into two types: tree-based and mesh-based. Tree-based peer-to-peer streaming systems build one or more trees rooted at the server. The media content flows from the server to the peers along the trees. Multiple trees are employed to make use of the uplink bandwidth of the leaf peers. Mesh-based peer-to-peer streaming systems build a loosely connected mesh where each peer establishes connection with several neighbours. Peers exchange the availability information of the media content with their neighbours and help each other to deliver the media content. Most of the works take little or no consideration of huge diversity of the link bandwidth of peers.

With the rapid expansion of the Internet, more and more individual computers are connected to the Internet which becomes potential peers for wireless streaming systems [8, 9]. Diverse and limited peer access link bandwidth is an important characteristic when these individual computers join a peer-to-peer streaming system. First, as different peers have different download bandwidths, it is desirable that peers with higher download bandwidth receive the media content at a higher bit rate while peers with lower download bandwidth receive the media content at a lower bit rate. This is achievable through the media streaming encoding technology (e.g., Multiple Description coding (MDC) and layered encoding). Besides, the overlay topology should guarantee a path from the source to the peer with sufficient bandwidth. Second, the limited uplink bandwidth should be utilized wisely so that the total downloading rate is maximized. In this paper, we propose a new scheme to optimize the overlay topology construction for peer-to-peer streaming systems with heterogeneous downloading requirements. Although the scheme is designed for wireless peer-to-peer media streaming systems, the result can also be applied to peer-to-peer content delivery or file downloading systems.

2. Overview of Proposed Communication System

Due to the NP hardness of the MDR problem, as a starting point, in this section we propose a greedy heuristic algorithm to find the maximum number of edge disjoint trees given the network topology G. It is a centralized algorithm which requires the complete information on the peers and their access link bandwidth. As we know, in practice, this information is dynamic and usually cannot be obtained in advance. We will further propose a distributed algorithm to handle the dynamics of peers in the next section. Nonetheless, the centralized heuristic algorithm serves as the foundation for the distributed algorithm and the benchmark when we evaluate the performance of the distributed algorithm in performance evaluation section.

The basic idea of the greedy heuristic algorithm is to pick out a maximum edge disjoint tree from graph G one by one until there is no such a tree. Before we apply the algorithm to graph G, we need to modify the graph by replacing each link of bandwidth b with b parallel links each of which has bandwidth 1. This modification will not change the result of the algorithm. Therefore, each tree represents a sub-stream with bit rate of 1. The receiving bit rate of a peer is equal to the number of trees the peer is on. Each tree must be rooted at the source node. To ensure the server transmits as many sub-streams as possible, the number of children of the root is limited to one for each tree. Two factors are taken into consideration for the tree construction. First, the height of the tree should be minimized. The height of the tree determines the delay between the source and the peer. To minimize the height of the tree, we should push the peers with higher uplink bandwidths to the source node as close as possible. Second, fairness bandwidth allocation between different trees should be maximized. When a peer receives the media content from several different trees, the delays to the source along the trees may be different. The final end-to-end delay from the source to the peer is determined by the worst case, i.e. the longest delay among all the delays. One way to minimize the difference among these delays is to minimize the difference of the tree heights, as the height of the tree represents the longest end-to-end delay approximately. Fairness here means that the uplink bandwidth should be shared among different trees in a fair manner. As a result, the height difference among different trees is minimized.

The fairness is realized by a property of the peer called fanout. Fanout is a value which is used to evaluate the forwarding ability of a peer. The fanout of a peer is equal to (uplink bandwidth)/(downlink bandwidth). The larger the fanout, the more children the peer can have. We use fanout as the upper bound on the number of children a peer can have in one tree. The reason is two folds. On one hand, if the number of children is smaller than the fanout, some uplink bandwidth will be wasted definitely even if the downlink bandwidth is filled up with the sub-streams. On the other hand, if the number of children is more than the fanout for one tree, it means that the number of children is less than the fanout for another tree. This imbalance of uplink bandwidth allocation contradicts the fairness principle.

3. Distributed Algorithm

The greedy heuristic algorithm works well if the information of the peers and their access link bandwidths are given in advance. However, in practice, peers join or leave the peer-to-peer system frequently which is known as *churn*. Due to the frequent churn rate, it is impossible for the server to run the greedy heuristic algorithm every time when there is a peer joining or leaving. When a new peer joins the system, it should be grafted to the system in an efficient and distributed fashion. The topology mismatch is another issue when we design a practical peer-to-peer streaming system. A practical peer-to-peer streaming system should be able to adapt itself to the changing overlay topology by alleviating the mismatch. As we mentioned before, peers join and leave the system in a random manner. According to the join procedure, a new peer is always a descendant of the existing peers after it joins the system. As a consequence, the overlay topology is changing as the peers join and leave and is dependent on the order of the peers joining and leaving. Due to the randomness of the order of peer joining and leaving, the overlay topology may be far from the one constructed by the greedy heuristic algorithm. We propose a topology adjustment procedure to help the peer-to-peer system handle the dynamic peer joining and leaving.

The intuition behind the adjustment procedure is to locate the peers whose positions do not match their level values and move these peers to a proper position. It is composed of three steps. (1) Step 1. The peer (We use v to denote the peer in the rest of this section) sends a REQUEST message upstream along the tree towards the root. The REQUEST message includes the value of level of v and a hop counter. Each peer receiving the REQUEST message will forward it to its parent and add 1 to the hop counter. The REQUEST message will stop when it reaches the root. (2) Step 2. Each peer receiving the REQUEST will compare the value of level in the message with its own value of level. If its own level is less than that in the message, it will send a GRANT message back to v. The GRANT message contains its own value of level, its parent, the residual uplink bandwidth of its parent and the value of the hop counter. (3) Step 3. If v does not receive any GRANT message, it does nothing. If v receives one or more GRANT messages, it will choose one peer to be its new parent based on the following rules: sort the peers in a descending order of their hop counts; select the first peer whose parent has non-zero residual uplink bandwidth and take its parent as the new parent of v. The peer v and the subtree rooted at v will become a subtree of the new parent. If there is no such a peer whose parent has non-zero residual uplink bandwidth, the peer checks itself to see if there is any residual uplink bandwidth. If no, the peer does nothing. If yes, the peer v will select the peer with the largest hop count and take its parent as the new parent. As there is no residual uplink bandwidth of the parent, one child of the parent will be replaced by v. The old child will become a child of v. The rationale behind the rules is to move the peer as close to the root as possible while minimizing the disturbance to the existing tree structure. Figure 1 shows the details of system model for material-based wireless data streaming.

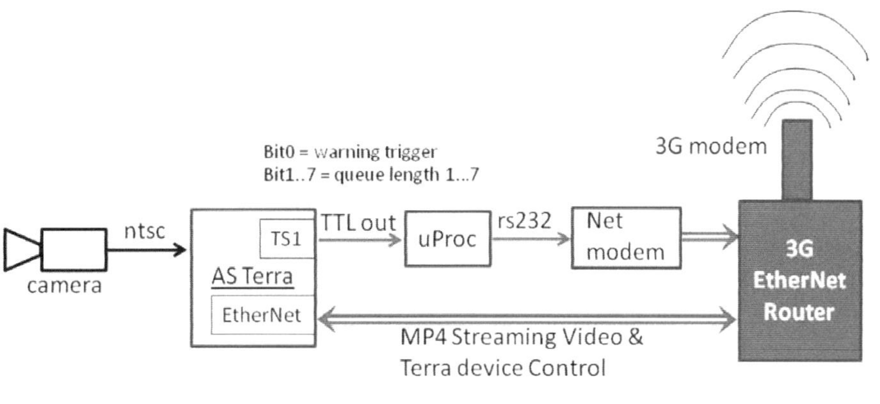

Figure 1. Basic system model of material-based wireless data streaming.

4. Performance Evaluation

In this section, we study the performance of the proposed algorithms through simulations. We have implemented the proposed scheme in NS-3. Peers are selected randomly from the stub networks and the bandwidth of the links in the transit network is set to be sufficiently high (1000 Mbps in the simulations). We compare our scheme with the Mitchell, Dem′ yanov and Malozemov (MDM) algorithm as it is the closest work to ours in the sense that both of the algorithms try to optimize the overlay topology for peer-to-peer media streaming systems. The main difference is that our scheme considers heterogeneous downloading rates while MDM assumes a uniform downloading rate. The simulation adopts three performance metrics such as satisfaction, end-to-end delay, and link stress.

We first compare the average satisfaction of peers under different sizes of the peer-to-peer streaming system. We use the number of peers in the system to represent the size of the system. Figure 2(a) shows the average satisfaction as we change the system size. We can observe that the greedy heuristic algorithm demonstrates the best satisfaction and the distributed algorithm outperforms MDM by about 30% on the average. It suggests that our scheme can make more efficient use of the uplink bandwidth of peers than MDM. With the increase of the system size, the average satisfaction drops for all the three schemes.

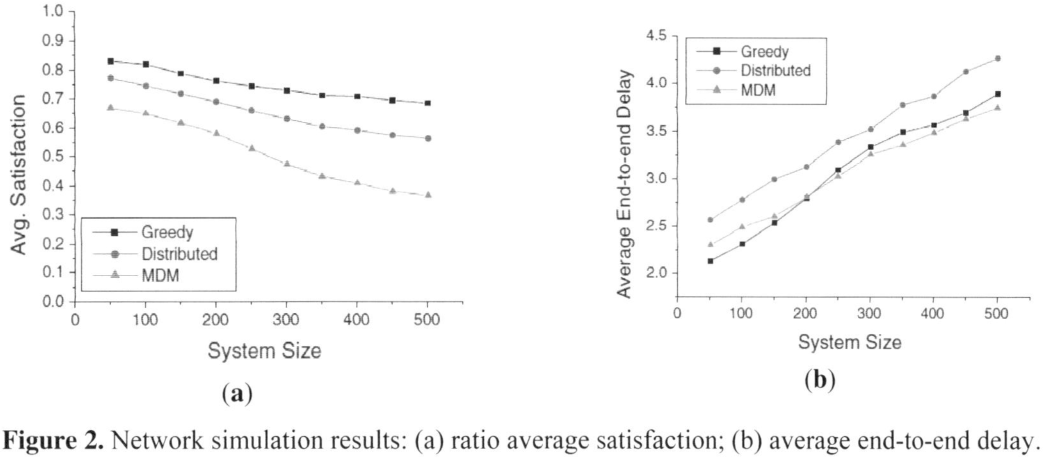

Figure 2. Network simulation results: (a) ratio average satisfaction; (b) average end-to-end delay.

The end-to-end delay is an important performance metric for peer-to-peer media streaming systems. As we can see later, sometimes end-to-end delay and satisfaction are two conflicting performance metrics and the best solution is a trade-off between these two metrics. Figure 2(b) shows the end-to-end delay under different system sizes. We can see that with the increase of the system size, the end-to-end delay increases as well due to the large tree heights. When the system size is small, the greedy heuristic algorithm achieves the shortest end-to-end delay. As the system size becomes large, the end-to-end delay of MDM is shorter than that of other two schemes. The reason is that MDM is a scheme focused on minimizing the end-to-end delay. The advantage of MDM is more obvious when the system size is large.

5. Conclusion

The Fourth Industrial Revolution fuses previous technologies that led to the transformation of industrial infrastructures and creates new technologies. A world in which things, things, things and people are connected and communicated organically. The world is the center of all technology. The core of Next Generation is convergence between ICT technologies [10, 11]. As material engineering progresses, the foundation for the growth of IT technology has been established. We have a plan to combine new materials with 5G communication technology.

References

[1] Uryu T, Jun Y and Yukio Y 2003 *J. Industr. Ecol.* Environmental fate of gallium arsenide semiconductor disposal: a case study of mobile phones, **7(2)** 103-12.

[2] Jo W, Landahi E C, Walko D A and Lee S 2018 *Appl. Physic. Lett.* Measuring femtometer lattice displacements driven by free carrier diffusion in a polycrystalline semiconductor using time-resolved x-ray scattering, **113(3)**.

[3] Kang B, Phuc N and Choo H 2018 *IEEE Sens. J.* Delay-Efficient Energy-Minimized Data Collection with Dynamic Traffic in WSNs, **18(7)** 3028-38.

[4] Kang B, Khuong Q and Choo H 2017 *PloS one* Implementation of fast handover for proxy mobile IPv6: Resolving out-of-order packets, **12(10)** 1-20.

[5] Kang B, Jo G and Choo H 2018 *J. Internet Technol.* A History-based Predictive Handover Scheme for Wireless Networks, **19(5)** 1487-97.

[6] Kang B and Choo H 2018 *Int. J. Sens. Networks* An energy-efficient routing scheme by using GPS information for wireless sensor networks, **26(2)** 136-43.

[7] Kang B, Myoung S and Choo H 2016 *IEEE Access* Distributed degree-based link scheduling for collision avoidance in wireless sensor networks, **4** 7452-68.

[8] Kang B and Choo H 2018 *J. Supercomput.* An SDN-enhanced load-balancing technique in the cloud system, **74(11)** 5706-29.

[9] Kang B and Choo H 2016 *EURASIP J. Wireless Commun. Networking* A cluster-based decentralized job dispatching for the large-scale cloud, **2016(25)** 1-8.

[10] Kang B and Choo H 2018 *ICT Express* An experimental study of a reliable IoT gateway, **4(3)** 130-3.

[11] Yeoum S, Kang B, Lee J and Choo H 2017 *Sensors* Channel and timeslot co-scheduling with minimal channel switching for data aggregation in MWSNs, **17(5)** 1-25.

Technology Research and Application on Catalyst Entire Life-cycle Management of Coal Fired Power Plant

Yang Zhang[1, 2], Qianwei Feng[1], Sha Li[1], Cunwen Li[1], Yukun Pei[1] and Yue Zhu[1]

[1]Huadian Electric Power Research Institute Co. Ltd., Hangzhou 310030, China
[2]State Key Laboratory of Clean Energy Utilization, Zhejiang University, Hangzhou 310027, China

cola1860@@foxmail.com

Abstract. Along with the comprehensive promotion of flue gas ultra-low emission reconstruction in coal-fired power plants, the stability, reliability and economic efficiency of SCR denitrification facility have emerged. SCR catalyst entire life-cycle management is the key to solving the problem. Given the present situation of the catalyst market, operation and maintenance, third-party service, a SCR catalyst entire life-cycle management technology have been proposed, including catalyst supply, quality control, operation and maintenance, effect optimization, life management and scrap disposal in the entire life-cycle of catalyst. By clarifying the accountability units at different stages, implementing corresponding quality control, operation and maintenance and evaluation and optimization technology, demonstration application suggest that the management technology could give full play to the professional expertise of participant companies, make the denitrification facility stable, reliable and economical, and bring obvious environmental and economic benefit.

1. Introduction

Since the 12th Five-Year Plan of China, large-scale denitrification and ultra-low emission renovation have been carried out in coal-fired power plants, among which SCR denitrification technology is the most widely used [1]. However, due to the centralized transformation task, lack of technical experience, inadequate quality control of major equipment, uneven quality of denitrification catalyst, and insufficient operation and maintenance experience of power generation enterprises, some hidden problems have gradually emerged after commissioning, such as catalyst wear, serious ash blocking, uneven distribution of flow field, performance indicators cannot reach the design value, air pre-heater blockage and corrosion are common problems [2]. In the current situation of coal-fired power plants promoting ultra-low emissions in an all-round way, higher requirements are put forward for the operation stability, reliability and economy of SCR denitrification unit. The key is to control the catalyst, which is the core of SCR denitrification.

Developed countries have experienced more than 40 years of industrial application of SCR denitrification technology. In view of many problems faced by SCR denitrification in the long-term operation process, a set of advanced and mature life-cycle management mode has been established [3-7]. However, compared with foreign countries, the quality of power coal in China is quite different, the supply of coal is unstable, the operation load is changeable, the operation mode and operation parameters of units are quite different, and the catalyst market has changed greatly in recent years, and the products are uneven. Therefore, it is not the best choice to apply foreign technology mechanically. Power generation groups, industrial companies and scientific research institutes have carried out a series of exploratory work in catalyst procurement and quality control, but the research and application of catalyst life-cycle management technology is still in its infancy, and mainly focus on guiding the operation and management of catalyst through third-party catalyst detection and life assessment [8-12]. Aiming at the problems existing in the application of denitrification catalyst, based on a lot of investigation and analysis, the technology of catalyst life-cycle management adapted to the present situation of denitrification application in China is studied and applied, and the product supply, quality control, efficient operation and maintenance, efficiency improvement optimization, life management and scrap disposal covering the whole life-cycle of catalyst are put forward. The life-cycle management model has been demonstrated and applied on a 1000MW unit, which can be used for reference and further discussion in the follow-up work.

2. Product Supply

Because the unit involves the transformation of ultra-low emission, the catalyst manufacturer is responsible for the supply of 404.8m3 catalyst with an additional layer, guarantees the performance of the catalyst throughout its chemical life cycle, replaces the damaged catalyst free of charge during operation, and disposes harmlessly of the spent catalyst with the same amount of loading after the expiration of the chemical life of the catalyst. Power generation enterprises adopt the payment method of 6:2:2 ratios in the chemical life of catalyst, and give the reward of 150,000 yuan/1000 h to the catalyst manufacturer after the catalyst life exceeds its life, which provides effective guarantee for the quality of catalyst supply.

3. Quality Control

Before the installation of the newly purchased catalyst, the third party technical service unit is responsible for sampling and testing the catalyst and evaluating the quality of the catalyst. As shown in table 1, the physical and chemical properties and process characteristics of the catalyst were tested by the third party technical service unit, and the comprehensive quality of the catalyst was evaluated according to the evaluation method agreed by the three parties [10]. The results showed that although some of the indicators were deviated from the reference values, the comprehensive quality of the catalyst could meet the requirements of use. At the same time, the test results were included in the database of catalyst performance parameters of the unit, which served as the basic parameters for the subsequent development of catalyst life management.

Table 1. Testing and evaluation results of new catalyst.

Category	Unit	Standard score	Reference value	Test value	Scoring method	Score
MoO_3	%	10	3	3.21	1 point will be deducted for every 0.1% lower	10
V_2O_5	%	10	3	1.33	2 points will be deducted for each 0.5% increase	10
SiO_2	%	5	6	7.06	1 point will be deducted for every 1% increase	3.9
Al_2O_3	%	5	2	1.95	1 point will be deducted for every 1% increase	5
SA	m^2/g	5	70	95.5	1 point will be deducted for every $2m^2/g$ lower	5
Adhesion strength	%	20	2.0	2.16	Level 1 is 20; Level 2 is 10; Level 3 is 5; Level 4 is 2; Level 5 is 0.	2
Wear strength	mg/100U	30	130	112	1 point will be deducted for every 1mg/100U increase	30
Wear strength	%	10	89.1	89.3	3 point will be deducted for every 0.5% lower	10
NH_3 slip concentration	mg/m^3		2.28	1.29	1 point will be deducted for every 0.05μL/L increase	
SO_2/SO_3 conversion rate	%	5	1 0.35	0.84 0.18	1 point will be deducted for every 0.05μL/L increase	5
Synthesis score					80.9	

4. Efficient Operation and Maintenance

Based on the operation and maintenance instructions of the catalyst manufacturer and the technical consultation of the third party technical service unit, the operation and maintenance personnel of the 1000MW unit strictly control the inlet flue gas parameters such as NOx concentration and dust concentration in their daily operation within the design scope, ensure the on-line instrumentation in a healthy state, and clean up the ash in time when the unit shuts down. Operators focus on the range of operating parameters such as NOx concentration, denitrification efficiency, ammonia injection rate, ammonia slip, flue gas temperature, catalyst pressure difference and air preheater pressure difference at the inlet and outlet of denitrification reactor, and deal with the abnormal parameters in time. Maintenance personnel focus on common problems of denitrification devices, such as blockage of ammonia supply valve, poor regulation performance, blockage of ammonia injection nozzle, deformation of diversion plate and rectifier device, deformation and failure of catalyst seals, ash accumulation and abrasion of catalyst, etc., and deal with them in time. Through the above efficient operation and maintenance work, the denitrification plant can be ensured to operate in a healthy and efficient state.

5. Efficiency Improvement and Optimization

In the process of ultra-low emission renovation of the 1000MW unit, the third-party technical service unit puts forward the renovation plan according to the present situation of denitrification unit. The catalyst manufacturer is responsible for the supply and performance assurance of 404.8m3 loaded catalyst. The module with serious wear and tear in the initial two-layer catalyst is replaced synchronously. The flow field is simulated and checked, and the top guide plate of the reactor is partial adjusted. As shown in table 2, the performance test results carried out by the third party technical service units after the transformation show that all performance indicators have reached the guaranteed value.

Table 2. Performance experiment results of the ultra-low emission retrofit.

Category	Unit	Guaranteed value	Test result	Remarks
Flue gas volume	m^3/h	3292467	3279829	
Inlet flue gas temperature	℃	356	372	
Inlet NOx concentration	mg/m^3	400	385	
Inlet dust concentration	mg/m^3	47000	30979	
Inlet SO_3 concentration	mg/m^3	28.8	24.1	
Flue gas temperature drop	℃	3	2	Qualified
Outlet NOx concentration	mg/m^3	-	27	
Outlet SO_3 concentration	mg/m^3	-	47.5	
NH_3 slip concentration	mg/m^3	2.28	1.59	Qualified
Denitration efficiency	%	87.5	89.8	Qualified
SO_2/SO_3 Conversion Rate	%	1.0	0.81	Qualified
System resistance	Pa	630		Qualified
A reactor	Pa		608	
Breactor	Pa		585	

6. Life Management

After running for one year, the 1000MW unit has some problems, such as lower denitrification efficiency, higher ammonia slip concentration, blockage of air preheater, etc. However, regular catalyst performance tests show that the denitrification efficiency meets the performance guarantee value, the ammonia escape rate is 1.38mg/m^3, and no abnormal attenuation of catalyst performance has been found. In order to solve this problem, the third party technical service unit carried out a performance evaluation and diagnosis test of the denitrification unit. According to the test results, the flow field of the denitrification unit was judged to have bigger problems. On this basis, the optimization test of denitrification and ammonia spraying was further carried out. As shown in figure 1, the non-uniformity of NOx concentration distribution at reactor outlet decreased from 42.0% and 44.5% to 10.0% and 10.1%, denitrification efficiency increased from 85.8% to 89.1%, and ammonia escape concentration decreased from 4.89 mg/m^3 to 1.76 mg/m^3, respectively. By carrying out the above work, the operation status of the denitrification unit was adjusted in time, and the problems such as increased consumption of denitrification ammonia and blockage of air preheater were solved, and the replacement of catalyst in advance was avoided.

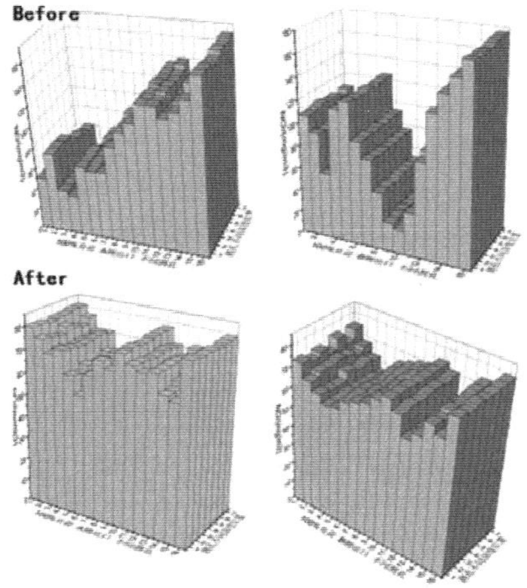

Figure 1. NOx concentration at the SCR reactor outlet before and after the AIG optimization.

On the premise of ensuring the denitrification efficiency, the catalyst life can be determined by testing whether the ammonia escape concentration exceeds the performance guarantee value of 2.28mg/m^3 [13]. As shown in figure 2, the catalyst test results and optimized test results show that the performance of catalyst can meet the application requirements. If the life of catalyst is determined according to the pre-optimization test results, the catalyst will be replaced in advance and the catalyst activity cannot be fully utilized.

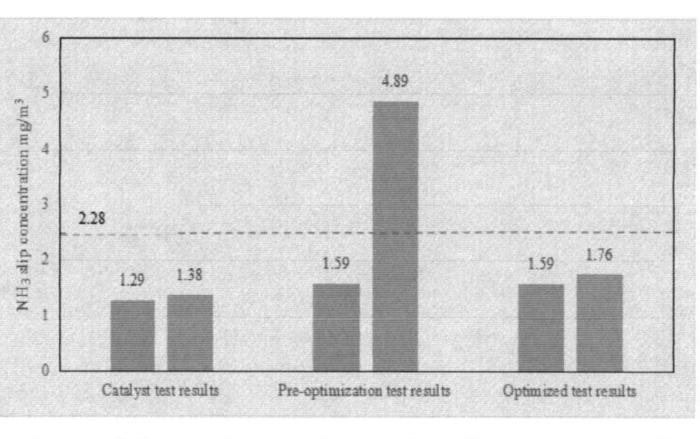

Figure 2. Analysis of the catalyst testing and performance experiment results.

7. Scrap Disposal

When the 1000MW unit purchases catalyst, it is agreed that the scrap catalyst will be recycled and disposed by the catalyst manufacturer. Considering that the catalyst purchased by the ultra-low emission modification will continue to be in service after the expiration of chemical life, the catalyst manufacturer is required to dispose of the catalyst with equal loading amount before replacing the catalyst [3, 4].

The demonstration application results of the above-mentioned 1000MW unit catalyst life-cycle management show that the implementation of life-cycle management can effectively enhance the NOx emission capacity, prolong the service life of the catalyst, reduce the disposal of scrap catalyst, reduce the energy consumption caused by the consumption of catalyst, and avoid the abnormal increase of operating resistance of denitrification unit and air preheater. The unit can reduce the accident, avoid the environmental risk caused by improper disposal of scrapcatalyst, so it can produce significant environmental and economic benefits.

8. Conclusion

Under the current situation that coal-fired power units are pushing forward ultra-low emission and environmental protection facilities into the stage of fine operation and maintenance, the existing catalyst management mode cannot meet the requirements of operation stability, reliability and economy of SCR denitrification unit, so it is imperative to carry out the life-cycle management of denitrification catalyst. The life-cycle management of catalysts should not only ensure the intensive cost of catalyst procurement, but also include the efficient performance control of denitrification plants. On this basis, the reliable, efficient and economical operation of denitrification plants can be realized. Therefore, the life-cycle management model of catalysts should cover the product supply, quality control, efficient operation and maintenance, efficiency improvement, life management and scrap disposal of catalyst throughout its life cycle. The six parts are not only connected in series, but also integrated with each other. The catalyst manufacturer is responsible for the product supply, efficiency improvement and waste disposal of the catalyst throughout its life cycle. The third-party technical service unit provides the technical service of catalyst quality control and catalyst life management. The power generation enterprise concentrates on the operation and maintenance of denitrification, which can give full play to the professional expertise of all units and has strong operability, and can produce remarkable environmental protection benefits and economic benefits.

References

[1] ZHU Y 2018 *Power Generation Technol.* Key technical issues of environmental protection island of thermal power plant, **39(1)** 1-12.
[2] FENG Q W, ZHANG Y, WANG F J, et al. 2017 *Electric Power* Analysis on operating status of SCR facilities in active coal-fired units, **50(4)** 157-61.
[3] Staudt J E and Engelmeyer A J 2003 SCR catalyst management strategies–modeling and experience, Combined Power Plant Air Pollutant Control Mega Symposium, Washington DC, 380-97.
[4] Staudt J E 2004 *Proc of ASME Power Conference. Batimore: ASME* Minimizing the impact of SCR catalyst on total generating cost through effective catalyst management, 731-40.
[5] Pritchard S 2006 *Power* Long-term catalyst health care, **150(1)** 1-5.
[6] Phananiramai P, Rosenberger J M, Chen V C P, et al. 2011 *Energy Syst.* A mathematical optimization technique for managing selective catalytic reduction for coal-fired power plants, **2(2)** 171-88.
[7] Nova I 2001 *Appl. Catalysis B: Environ.* Study of thermal deactivation of a de-NOx commercial catalyst, **35(12)** 31-42.
[8] YAO Y, YANG X N, Kong F H, et al. 2017 *Proceedings of the CSEE* Performance analysis and evaluation of honeycomb SCR catalysts in coal-fired power plant, **37(S)** 112-6.
[9] SONG Y B, YANG J, JIN L P, et al. 2016 *Electric Power* Study on methodology of SCR catalyst macroscopical performance evaluation and lifetime prediction, **49(4)** 17-22.
[10] DU Z, YANG L Q, YAN M, et al. 2016 *Electric Power* Construction and implementation of the whole process management for SCR denitration catalyst, **49(4)** 12-6.
[11] FU Y, LU Q, TANG S J, et al. *Electric Power* 2018 Study on life management of the SCR denitrification catalyst, **51(3)** 163-9.
[12] FU Y, LU Q, ZHUANG K, et al. 2017 *Electric Power* Life prediction for SCR flue gas denitrification catalyst in coal-fired power plants, **46(7)** 60-5.
[13] FANG C J, LU C Z and BAI X L 2018 *Electric Power* Study on operation optimization of SCR denitrification with ammonia-injection for ultra-low emission, **51(2)** 143-8.

Application Research on Magnetic Coupling Stress Testing Technique for In-service Steel Bridge

WENG Guangyuan[1], SHI Yun[1], ZHANG Yumin[1] and DAI Jianbo[1]

[1]Mechanical Engineering College, Xi'an Shiyou University, Xi'an 710065, China

shiyun0305@163.com, weng_guangyuan@163.com, zhangyumincorn@163.com, d_jianbo@foxmail.com

Abstract. Magnetic effect of ferromagnetic materials was discussed, the magneto-mechanical properties of ferromagnetic materials were studied, as well, Jiles Atherton constitutive mode was studied. Through the establishment of concrete filled steel tubular arch bridge structure calculation model, analysis the boom in tension and compression of the magnetic coupling. Results show that the resultant magnetic coupling model of Q345 steel derrick, the boom and load changes, the change of the magnetic parameters is more sensitive, can be better applied to the stress testing of such components.

1. Introduction

In recent years, with the development of transportation, the steel structure bridge has been widely used. But due to the stress of the steel bridge structure are complicated, it is difficult for designers to accurately estimate the actual internal force values of each link and node. Therefore, in order to detect stress of the steel bridge structure, and, in order to assess the reliability of steel bridge in service, the changes rules between the design internal force and the actual value have been studied, nondestructive stress testing for steel bridge structure is one of the hot issues, stress detection based on magnetic coupling effect got the attention of the scholars both at home and abroad [1-2]. In this paper, take Suspenders of CFST arch bridge as the research object, the principle of magnetic coupling stress testing technology is researched from a new Angle.

2. Maginetic Effect of Ferromagintic Materials and Magnetic Hysteresis Model

2.1. Magnetic Effect of Ferromagnetic Materials

Magnetic effect refers to the ferromagnetic materials is associated with spontaneous magnetization and technology magnetization between magnetic and mechanical properties of the effect of mutual contact and influence each other.

Magnetic effect mainly includes the following several aspects

(1) the magnetostriction

Magnetostrictive effect is the most common and most widely used magnetic effect [3-4]. Commonly used magnetostrictive effect refers the ferromagnetic materials when subjected to external magnetic field, the tiny changes of the length of the material in magnetic field direction. The rate of change is called magnetostrictive coefficient of per unit length, for ferromagnetic materials, average between 10-6-10-5.

(2) magnetic volume effect

This effect refers the outer ferromagnetic materials when the magnetization of the magnetic field, its volume change. Because of this effect first observed in the metal Ni by Barrett, it is also known as Barrett (Barrett) effect. Magnetic volume effect in ferromagnetic materials from paramagnetic state after the Curie temperature into a sequence of magnetic state will have the volume change, this is due to the effect of exchange of the ferromagnetic material in magnetic ordered field caused by the spontaneous magnetization, rather than caused by the technology of magnetization of the magnetic field outside, sometimes it is called exchange magnetostriction. Magnetic inverse effect of volume effect is also called the Nagaoku-Honda effect.

(3) magnetic torsion effect

This effect refers to the proper magnetic circuit is formed on the magnet, when a current through the magnetic body distortion will occur and can be used to make reverse motor, etc. This effect is also known as Wiedenmann effect, and its inverse effect refers to a longitudinal magnetization of ferromagnetic rod exert torque, the causes of longitudinal current change, namely that the circumference of the vertical rod axis direction has a magnetic field and the change of intensity of magnetization, also known as Wertheim effect. This mechanical distortion in the magnetic body, and in the secondary coil current AnTi-Wiedenmann effect, can be used in the production of all kinds of torsion sensor.

(4) magnetic bending effect

This effect is a ferromagnetic material bar bending, and when the vertical (axial) magnetization, the phenomenon of iron bar would be straightened, the magnetic field makes a ferromagnetic material rod bending effect, also known as Guilleanin effect, can be used in the production of all kinds of curved actuators.

(5) magnetic rotation effect

This effect refers to the free hanging in the direction of the axis of the ferromagnet with external magnetic field or the external magnetic field is reversed [5], ferromagnetic experience by torque effect and produce the effect of the rotation, also known as Einstein DE Haas effect. Its inverse effect called Barnett effect, which is when the original not magnetized ferromagnetic turns, will produce certain magnetization. Magnetic rotation effect associated with ferromagnetic micro electronic spin in the body.

(6) magnetic elastic effect

This effect refers to the ferromagnetic material under plus a magnetic field, the modulus of elasticity (young's modulus) will change effect, also known as the ΔE effect or Brackett effect, can be used for acoustic delay line or sensors [6].

(7) magnetic rigid deformation effect

This effect refers to the ferromagnetic material under plus a magnetic field, its stiffness coefficient will change effect, also known as the Kimball effect, which can be used in the production of sensors or actuators.

2.2. Magnetic Hysteresis Model

Assume that polycrystalline ferromagnetic materials for isotropic materials, under the effect of magnetic field H, its typical magnetic domain (with the magnetic moment m), per unit volume energy type as follows

$$E = -\mu_0 m \cdot H \tag{1}$$

In equation (1), H is within the material within the practical domain under magnetic field intensity, not the external magnetic field.

The coupling effect between the introduction of magnetic domain, the energy per unit volume can be expressed as

$$E = -\mu_0 m \cdot (H + \alpha M) \tag{2}$$

In equation (2), α is an average field coupling parameters between adjacent magnetic domain.

Set up effective magnetic field $H_e = H + \alpha M$, the magnetization of the response of the effective field can also be expressed as follows

$$M = M_s f(H_e) \tag{3}$$

In equation (3), f for arbitrary function of effective field; At that time $H_e = 0$, $f(0) = 0$. At that time $H_e \to \infty$, $f(\infty) = 1$. M_s for saturated magnetization.

Formula (3) and the corresponding to the lowest energy state of the magnetic domain statistical distribution, and does not take into account related to material structure characteristics, such as location and nonmagnetic impurity inclusions, etc. Therefore, using this equation can be obtained without hysteresis magnetization curve or ideal magnetization curve. Under the effect of magnetic field H, the domain walls along the curve to reach real equilibrium position. Therefore, the equation can be written as the type

$$M_{an}(H_e) = M_s f(H_e) \tag{4}$$

In equation (4), M_{an} for no hysteresis magnetization.

For common no hysteresis magnetization, selection function f for the Langevin function $L(H_e)$, available

$$M_{an}(H_e) = M_s(\coth(H_e/a) - (a/H_e)) \tag{5}$$

In equation (5), a has the dimension of magnetic field intensity, characterization of no hysteresis shape of the curve, and is proportional to the effective density of the magnetic domain and absolute temperature, can be defined as follows $a = k_B T / \mu_0 \langle m \rangle$, among them, k_B for the Boltzmann constant (1.38×10^{-23}jk^{-1}), T is the absolute temperature, μ_0 for the vacum magnetic permeability ($4\pi\times10^{-7}$ H/m), $\langle m \rangle$ for effective magnetic domain size. Revised Langevin equation does not describe the normal dc magnetization of ferromagnetic well, because the model ignores the magnetization of the irreversible changes, such as when the domain wall movement to pinning point location, the magnetization of the irreversible change.

When the magnetic field does total energy per unit volume is equal to the balance of specimen magnetic energy required to do the work of a magnetic field, the magnetic domain wall will return to its starting position, namely

$$E = \int HdB = \frac{1}{\mu_0}\int BdB - \int MdB \tag{6}$$

Among them, the external magnetic field on the right side of the second said on the specimen of the work done.

When there is no pinning point of ferromagnetic domain wall is pressure, moving in the domain wall, magnetization along no hysteresis curve to reach equilibrium. Therefore, the equation can be expressed as follows

$$\int M dB_e = \int M_{an}(H_e) dB_e \tag{7}$$

In equation (7), $B_e = \mu_0 H_e$

It is well known that under the action of outside magnetic field, by pinning point in ferromagnetic materials (such as non magnetic inclusions, voids, non-uniform stress zone), have set back domain wall movement. The existence of these defects will lead to the decrease of the initial magnetic permeability and ferromagnetic materials with the increase of coercive field.

3. Analyses of Numerical

3.1. Model Structure

This project depends on the Lan-cang river bridge construction project, located in the area of Tibet Changdu town, the basic intensity of Ⅶ degrees in this region, the ground motion peak acceleration of 0.1 g, the design, the bridge seismic press Ⅷ degrees fortification, main bridge upper structure using single span 125 m, half through bridge concrete filled steel tubular arch bridge with 31.25 m rise, arch axis equation of catenary, arch axis coefficient is 1.347, for precast reinforced concrete hollow slab bridge and arch bridge using bored piles foundation. Little pile number approach of upper with 15 m prestressed reinforced concrete hollow slab, approach of using in the pile size 1-15 m prestressed concrete hollow slab and 1-15 m sector plate.

In five main arch rib section of production, arch rib dimensions are subject to loft to arch rib arch axis, namely for the arch rib axis length L, arch rib steel pipes of longitudinal seam, the butt weld require the use of automatic welding, full penetration, groove form is determined by the welding way; All welds are Ⅰ weld.

The arch bridge of the arch rib in the 2.7 m high, arch rib on the bottom chord pole diameter of 950 mm, adopt 20 mm thick steel plate. Bridge upper structure of concrete filled steel tube adopts C30 concrete and prestressed concrete beams and surfacing layer USES the C40 concrete, adopting C35 concrete bridge panel.

3.2. Establish the Model Structure

The calculation model was established based on general finite element calculation program structure, respectively, set up concrete filled steel tube concrete material for ordinary concrete and different replacement rate of recycled concrete structure calculation model. The calculation model of the arch structure as shown in figure 1.

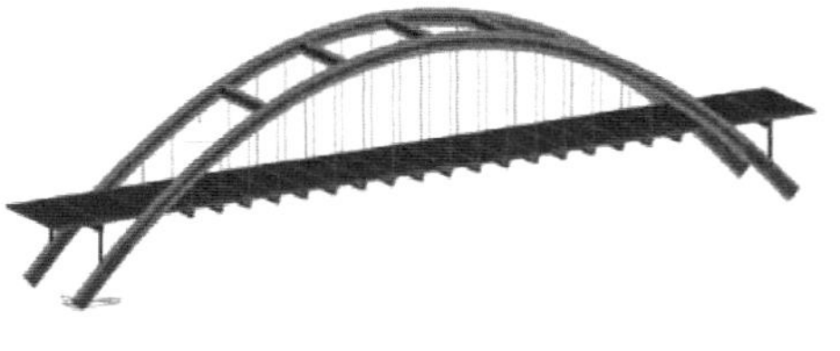

Figure 1. Model structure.

A single arch rib is divided into 24 section, section forms of dumbbell arch section, the end section of the arch foot I combination section, the section form and cross section geometric parameters as shown in figure 2(a), in addition to the arch foot end section, the rest of arch ring cross section Ⅱ combined section, the section form and the cross section of geometrical parameters as shown in figure 2(b).

(a) combination section I section Ⅱ (b) combination

Figure 2. Dumbbell arch cross section.

In order to simplify the calculation, the mechanics model is set up, according to the area and the principle of changeless of section steel content and the section of equivalent to the above two kinds of ideal round steel tube concrete section, simplified section geometric parameters and steel content are shown in table 1.

Table 1. Simplified model structure section.

cross section features	cross section type	
	combination section I	combination section Ⅱ
containing steel ratio (%)	9.2	8.1
cross section diameter D (mm)	400	380
thickness of the steel tube t (mm)	2.80	2.76

3.3. Technical Solution

This paper intends to combine the theoretical analysis and computer simulation analysis method, the deformation and stress of ferromagnetic materials can be expressed as the constitutive relationship between magnetic field.

$$\begin{aligned} \varepsilon_{ij} &= \varepsilon_{ij}(\sigma_{ij}, H_i) \\ B_i &= B_i(\sigma_{ij}, H_i) \end{aligned} \tag{8}$$

In equation (8), σ_{ij} as the stress, ε_{ij} strain, B_i as the magnetic induction intensity, H_i magnetic field intensity. It can be seen that when the independent variable is stress and magnetic field intensity, the output is the strain and magnetic induction intensity, so it gets in the experiment are must measure the amount of basic, test is difficult. In the applied load and magnetic field, under the condition of measuring the specimen deformation and magnetic induction intensity, reflected by magnetic force characteristic curve of ferromagnetic materials, such as magnetic hysteresis and magnetostriction curves and magnetic induction intensity – stress curves, etc., at the same time be able to reflect the magnetomechanical properties of ferromagnetic materials important parameters, lay the foundation for the theoretical research and engineering application.

3.4. The Calculation Results

Based on the vault boom, for example, set in the calculation of derrick for Q345 steel, the vault derrick and magnetic loading at the same time, the magnetic field is obtained by numerical simulation calculation parameters and the corresponding relation of derrick stress. Under the action of tensile stress in the derrick model of magnetization curve of the specimens is shown in figure 3. Can be seen from the diagram, when the magnetic field intensity, namely when the magnetization process of transition from the second to the third phase, the stress sensitivity of magnetization gradually, and with the increase of tensile stress, the magnetization and increasing, related close to the linear change.

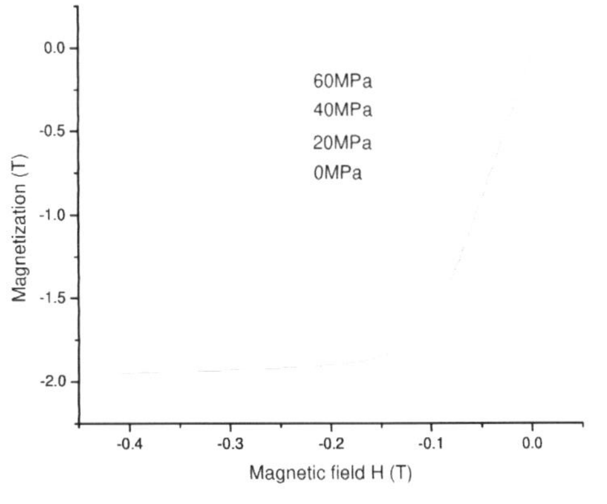

Figure 3. The magnetization curve for specimens under the action of tensile stress.

When to compressive stress, compressive stress test magnetization curve of the specimens under the action of the derrick model as shown in figure 4. Can be seen from the diagram, when the magnetic field intensity, namely when the magnetization process of transition from the second to the third phase, the stress sensitivity of magnetization gradually, and with the increasing of compressive stress and the intensity of magnetization is constantly decreases, and still close to the linear relationship between related changes.

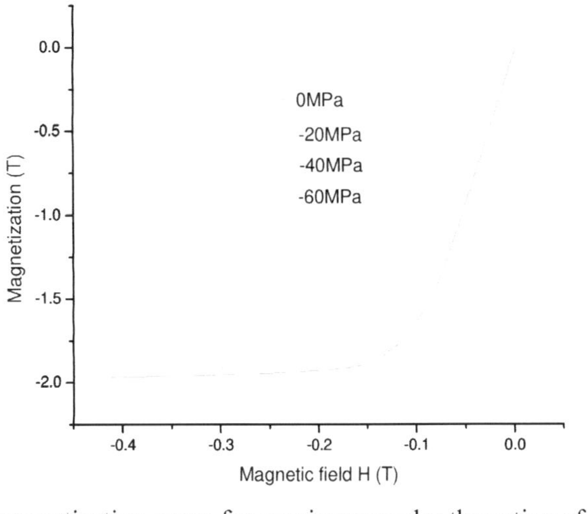

Figure 4. the magnetization curve for specimens under the action of compressive.

IOP Conf. Series: Materials Science and Engineering **649** (2019) 012030 doi:10.1088/1757-899X/649/1/012030

Conclusions

Various magnetic effect of ferromagnetic materials is discussed in this paper, study the magnetomechanical properties of ferromagnetic materials, studies the Jiles Atherton constitutive model. Through the establishment of concrete filled steel tubular arch bridge structure calculation model, analysis the boom in tension and compression of the magnetic coupling. Results show that the resultant magnetic coupling model of Q345 steel derrick, the boom and load changes, the change of the magnetic parameters is more sensitive, can be better applied to the stress testing of such components.

Reference

[1] Maugin G A and Sabir M 1990 *Int. J. Plasticity* Mechanical and magnetic hardening of ferromagnetic bodies: Influence of residual stresses and applications to nondestructive testing, **6(5)** 573-89.

[2] Gilanyi A, Morishita K, Sukegawa T, et al. 1998 *Fusion Eng. Design* Magnetic nondestructive evaluation of fatigue damage of ferromagnetic steels for nuclear fusion energy system, **42** 485-91.

[3] Doubov A A and Demine E A 2000 Problems of welded joints quality control and possibility of metalmagnetic memory method using, 15th World Conference 15-21 October 2000 in Rome.

[4] Dorbof A A and Demin E A 2001 *Appl. Phys.* Application of the metal magnetic memory and computer Devices for contol of welding quality, **2**.

[5] Banan M R and Hjelmstad K D 2015 *J. Struct. Eng.* Parameter estimation of structures from static response. I. Computational aspects, **120(11)** 3243-58.

[6] Devriendt C, Weijtjens W, De Sitter G, et al. 2013 *Mech. Syst. Signal Process.* Combining multiple single-reference transmissibility functions in a unique matrix formulation for operational modal analysis, **40(1)** 278-87.

Acknowledgements

The work was funded by Key Research and Development Program of Shaanxi (Program No.2019SF-266), Science Project of YuLin government (Grant No. 2016-29), Shaanxi Natural Science Foundation (Grant No. 2018JQ5184, 2018JQ5073) and National Natural Science Foundation of China ((Grant No.51808446).

Seismic Effect Research of Seismic Measures for Beam Bridge under Different Earthquakes

Yu-min ZHANG[1], Guang-yuan WENG[1], Yun SHI[1] and Jian-bo DAI[1]

[1]Mechanical Engineering College, Xi'an Shiyou University, Xi'an 710065, China

zhangyumincorn@163.com, weng_guangyuan@163.com, shiyun0305@163.com, d_jianbo@foxmail.com

Abstract. Collision, collapse and expansion device damage are very likely to occur in bridge structures under large earthquakes. Earthquake waves with exceeding probability of 63.2%, 10%, and 2% in 100 years, as well as the Northridge earthquake and Kobe earthquake have been used to analyze the seismic response under earthquakes. The result shows that collision and collapse damage of bridge can be effectively controlled by the seismic measures. When the peak acceleration of earthquake is large and the control ability of the limit device to the excessive displacement of the structure is insufficient, the unseating prevention device can effectively make up the deficiency of the limit device to limit the excessive displacement of bridge.

1. Introduction

In recent years, the unseating of bridge superstructures after collisions and explosions have been widely concerned by the society. As the superstructure is simply supported on the top of the substructure, the beam bridges were more likely collapse under the impact load.

There were many bridges falling down from piers and abutment in the San Fernando earthquake. In 1989, the approach bridge of Auckland Bridge falling off from substructure because the supporting surface was too narrow. Unseating of bridges has become a common earthquake damage in many large earthquakes. It is found that the failure of displacement limiting measures and the insufficient width of superstructure are the main reason of collapse damage, as well the collision damage can hardly be avoided in the earthquake [1].

In order to prevent collision and unseating damage of bridge under earthquakes, many scholars are devoted to studying the control measures of such earthquake damage. The seismic design code of bridge is revised, and it is suggested that when the intensity is greater than 8 degrees, the seismic measures to limit the excessive dispplacements of bridge should be adopted [2].

2. Damage-Reduction Seismic System of Bridge

As damage-reduction seismic design to be used widely in bridge seismic design, many seismic measures have been applied at the end of bridge beam, such as the limiting device and unseating prevention devices. The measures are used to limit the excessive displacement of the structure under earthquake, and are special design to achieve different seismic requirements and functions [3].

The limit device is mainly used to limit major displacement of bridge under earthquake, in order to prevent the failure of bearing. It is found that the relative displacement between adjacent beams as well as between superstructure and substructure at beam end can be effectively reduced by installing tensile limiting device. But the demand for the ductility of piers at edge span is increased [4]. Unseating prevention devices is effectively to prevent collision and unseating damage of bridges, as bearings are damaged or the limit device is insufficient to control the excessive displacement of bridge under severe earthquakes [5].

Although researchers have studied separately on the seismic performance and the limit effect of limit device and unseating prevention devices, the studies on control effect of damage-reduction seismic system consisted by limit device and unseating prevention devices were few, especially for the collision and unseating damage in rare or unexpected earthquakes [6]. The bridge with different combinations of damage-reduction seismic measures will be analyzed to obtain the limits and seismic effects of various seismic measures.

3. Finite Element Analysis Model

A continuous beam bridge with 5 spans of 30m has been analyzed, the box girder of bridge is 17m wide and 1.6m high with concrete C50. The laminated rubber bearing were been used to support the superstructure. The piers are 30 meters high and the diameter are 2.0 meters, and the concrete C30 is adopted. The finite-element model of bridge was shown in figure 1.

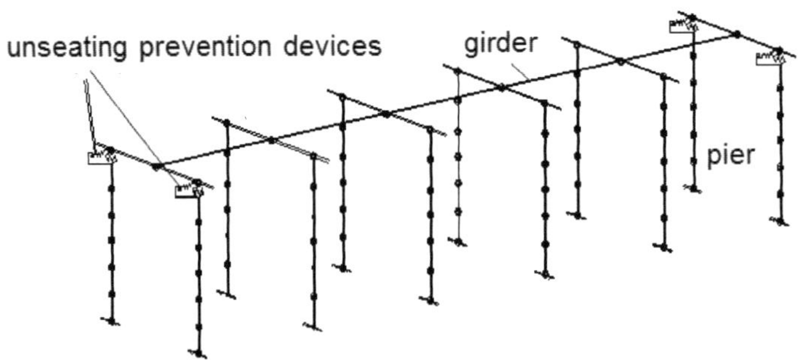

Figure 1. Finite-element model of bridge.

The girder, capping beam and piers were all simulated as beam elements in the finite element model of the bridge. The bearings were simulated by elastic connection without considering the influence of pile foundation on bridge structure.

In order to analyze the seismic response of bridge and the application effect of seismic measures, the seismic loads adopted in bridge are increased step by step. There are three earthquake waves with exceeding probability of 63.2%, 10%, and 2% in 100 years (small, moderate, and large earthquakes), as well as the Northridge earthquake in the United States and Kobe earthquake in Japan. The peak acceleration of these seismic waves shows an increasing trend, as shown in table 1.

Table 1. The peak acceleration of design earthquakes (cm/s^2).

Name	Small earthquake	Moderate earthquake	Large earthquake	Northridge earthquake	Kobe earthquake
peak acceleration	96.20	250.15	363.90	592.08	617.14

The time history of the Kobe earthquake wave was shown in figure 2.

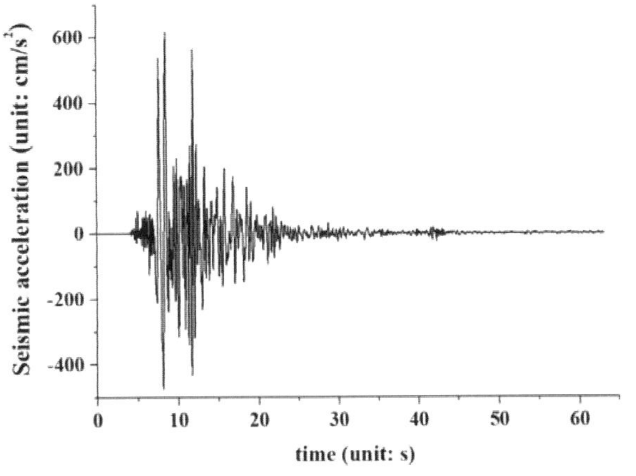

Figure 2. Time history of Kobe Earthquake.

4. Seismic Response of Bridge under Different Earthquakes

The seismic response of the bridge can be obtained by using the nonlinear time-history analysis under the excitation of different earthquakes. The seismic analysis results have shown in table 2.

Table 2. Seismic responses of bridge under different earthquakes.

Earthquakes	Positions	Beam displacement (cm)	Displacement of pier top (cm)	Bearing displacem ent (cm)	Bending moment of pier bottom (kN·m)
small	Side pier	3.92	0.21	3.71	1374
earthquake	middle pier	3.92	0.91	3.01	5373
moderate	Side pier	14.99	0.59	14.40	3858
earthquake	middle pier	14.99	3.25	11.74	19108
large	Side pier	19.12	0.78	18.34	5099
earthquake	middle pier	19.12	4.28	14.84	25258
Northridge	Side pier	24.92	1.14	23.78	7564
earthquake	middle pier	24.92	5.54	19.38	32594
Kobe	Side pier	19.06	1.38	17.68	9047
earthquake	middle pier	19.06	4.25	14.81	33742

As the results been shown in table 2, the seismic responses such as beam displacement, displacement of pier top, bearing displacement and bending moment of pier bottom increased gradually as the earthquake intensity increased, and the top displacement and bottom bending moment of middle pier are larger than those of the side piers.

It must pay attention to the displacement of bridge, the displacement of beam-end will exceed the permissible displacement of the expansion device as the peak acceleration of the seismic load exceeds moderate earthquake, and the bearing displacement have exceeded its shear deformation. This means that bearing and expansion device will be destroyed as the earthquake intensity larger than moderate earthquake, the beam of bridge may be falling down from piers or to be collision under larger earthquake. So it is necessary to adopt seismic measures to limit its excessive displacement.

5. Seismic Response of Bridge Setting on Seismic Measures

In order to obtain the influence rule of seismic measures on the seismic response of bridge under the action of earthquake, the original design structure with different combination seismic measures is defined as different cases for comparative study. The distribution of cases is as follows:

Case 1: original design structure;

IOP Conf. Series: Materials Science and Engineering **649** (2019) 012031 doi:10.1088/1757-899X/649/1/012031

Case 2: original design structure + expansion device;

Case 3: original design structure + expansion device + limit device;

Case 4: original design structure + expansion device + limit device + unseating prevention device.

The results show that seismic measures have obvious influence on seismic response of bridge, especially the beam end displacement and bearing displacement of side piers. The variation of the seismic response of the structure setting on different combinations of seismic measures are shown in figure 3 and figure 4.

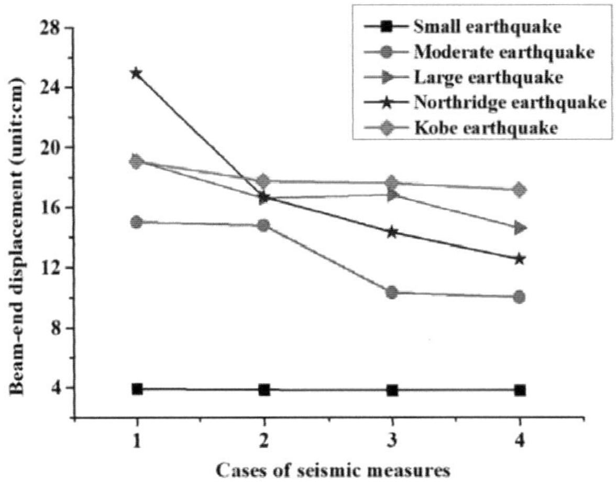

Figure 3. The displacement of beam-end under earthquakes with different case of seismic measures.

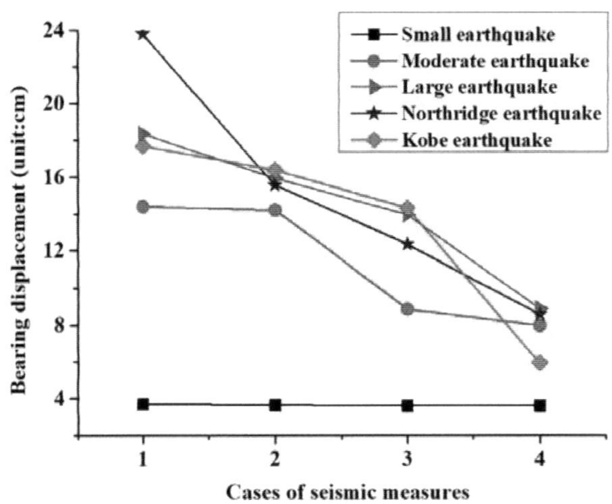

Figure 4. Bearing displacement of side piers under earthquakes with different case of seismic Measures.

As shown in figure 3 and figure 4, the displacements of the beam-end and bearings under earthquakes have greatly been decreased as seismic measures to be adopted, except for small earthquakes excited. And all seismic measures have play a role to decrease the displacement of bridge, according to the changing trend of the curve, the displacement of beam-end and bearings of the side piers were both decrease with the increase of seismic measures, and the bearing displacement of the side piers decreases greater than middle piers.

As limit device and the unseating prevention device both adopted on the bridge (case 4), the displacement of the beam-end (at the expansion joint) was almost controlled within the permissible displacement of the expansion device under the action of various earthquakes, and the collision damage could be avoiding except for under the largest Kobe earthquake. The displacement of side piers can be well controlled to prevent the superstructure falling off from the substructure.

In case 4, the limit device and the unseating prevention device were combined use to limit the displacement of bridge under earthquakes, the limit device come into play as the relative displacement of superstructure and substructure lager than 3.2 cm, when the relative displacement increases to 8 cm the unseating prevention device will in action. The stress of the limit device was decreased as the unseating prevention device working with the limit device.

6. Conclusion

In view of the collision and collapse of bridge under large earthquake, combined with the existing research conclusions of seismic measures such as limit device and unseating prevention device, the nonlinear time-history analysis of bridge with different seismic measures have been carried under different earthquakes. The main conclusions are as follows:

Collision, collapse and expansion device damage are very likely to occur in bridge structures under large earthquakes. The collision and beam falling damage of bridge can be effectively controlled by the seismic measures such as limit device and unseating prevention device.

For the bridge structure considering seismic design, the displacement demand of expansion device under earthquake should be considered, expansion device with large displacement should be selected. The limit device can effectively reduce the displacement of beam and bearing, and can also increase the overall stiffness of the bridge. However, the limiting capacity of the limiting device is slightly insufficient under larger earthquake.

When the peak acceleration of earthquake is large and the control ability of the limit device to the excessive displacement of the structure is insufficient, the unseating prevention device can effectively make up the deficiency of the limit device to limit the excessive displacement of bridge. The limit device and unseating prevention device can form an optimized limit system, which can reasonably distribute the impact force of earthquake, so as to ensure the integrity of the structure.

References

[1] Padgett J E and Desroches R 2008 *Eng. Struct.* Three-dimensional nonlinear seismic performance evaluation of retrofit measures for typical steel girder bridges, **30(7)** 1869-78.

[2] Ruting B 2012 *J. Earthquake Eng. Eng. Vib.* Analysis of seismic failure mechanism and earthquake resistant measures for the abutments of Gaoyuan bridge, **32(4)** 79-87.

[3] Andrawes B and Desroches R 2005 *Smart Mater. Struct.* Unseating prevention for multiple frame bridges using superelastic devices, **14(3)** 60-7.

[4] Yan B and Dai G 2013 *Struct. Eng. Int.* Seismic Pounding and Protection Measures of Simply-Supported Beams Considering Interaction between Continuously Welded Rail and Bridge, **23(1)** 61-7.

[5] Zhang Y M and Qin X X 2014 *Appl. Mech. Mater.* Research on Bridge Damage-Reduction Seismic Performance with Variable Parameter Cable-Unseating Prevention Devices, **638-640** 977-82.

[6] Zhang Y 2018 *IOP Conf. Series: Earth Environ. Sci.* Study on seismic behavior for high damping rubber bearings of continuous beam bridges, **153(5)** 1-5.

Acknowledgements

The work was funded by Shaanxi Natural Science Foundation (No. 2018JQ5184, 2018JQ5073), Science Project of YuLin government (No. 2016-29) and National Natural Science Foundation of China ((No.51808446).

Institute of Physics
Dirac House, Temple Back
Bristol BS1 6BE UK

ISSN: 1757-8981
ISBN 978-1-5108-9846-2

CURRAN ASSOCIATES INC.
proceedings.com